初级人体解剖图谱
ATLAS OF HUMAN ANATOMY
FOR JUNIOR PREMEDICAL STUDENTS

徐存一　著
(Cunyi George Xu)

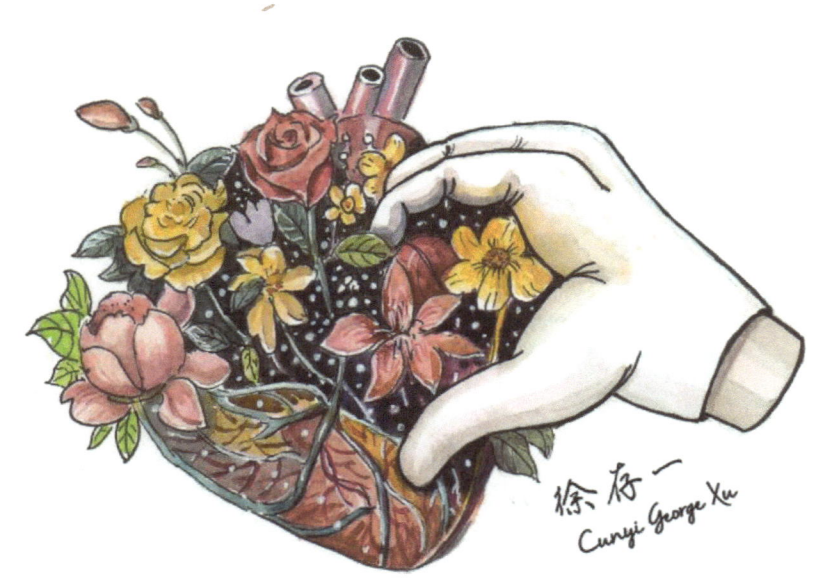

徐存一
Cunyi George Xu

Billson International Ltd.

自序

奈特图谱是弗兰克奈特先生（Frank H. Netter）手绘的《Netter's Atlas of Human Anatomy》（《奈特人体解剖图谱》），是一本经典的医学解剖学图谱，包含了大量详细的人体解剖插图，广泛用于医学教学和学习。书中的插图详细展示了人体解剖结构，包括骨骼、肌肉、神经、血管等，帮助医学生、医生和其他医疗专业人士更好地理解人体解剖，但是由于这本图谱包含了大量的专业词汇，以及解剖层面的医学细节，对于高中生来说，这本书太难了。同时现在发行的大多数科普类的医学绘本，对于高中生来说，又过于简单，而且缺乏专业性。

我作为一个正在读 A-Level 的学生，从我能理解的视角，拟出版一系列能够适合高中生的医学图谱《Atlas of Human Anatomy for Junior Premedical Students》，旨在提高高中生对人体解剖知识和细胞功能和结构的认识，希望本书能提供过渡阶段的帮助，为高中生申请未来在大学就读医学院或者生物学提供帮助。也希望这本书的出版能够吸引更多热爱绘画和将要申请医学院的高中生共同完成一系列的作品集。图书一旦出版，希望能得到更多专业医学专家的指导。

本书所有图片为作者手绘，扫描，然后完成标注，并添加简单功能注释。为了能引起更多的读者兴趣，这个绘本提供了中英双语的标注。本书所获得利润将全部捐赠，用于医学相关的教育。

Author's Preface

The *Atlas* is a hand-drawn collection by Mr. Frank H. Netter, titled *Netter's Atlas of Human Anatomy*. It is a classic medical atlas of human anatomy widely used for medical education and learning. This atlas contains numerous detailed anatomical illustrations, showcasing the structure of bones, muscles, nerves, blood vessels, and more. It aids medical students, doctors, and other healthcare professionals in better understanding human anatomy. However, due to the large number of unfamiliar medical terms and anatomical details, this atlas can be too difficult for high school students to understand. On the other hand, most of the science-focused medical popularisation picture books currently available on the market are overly simplified for high school students and lack the necessary professionalism.

As an A-Level student, I plan to publish a series of medical atlases titled *Atlas of Human Anatomy for Junior Premedical Students* from a perspective that I, as a student, can understand. The purpose is to enhance high school students' knowledge of human anatomy, cellular functions, and structures. I hope this picture book will serve as a helpful transition for those students preparing to apply to medical school or biology-related subjects in their university studies. Additionally, I aspire for this publication to attract more high school students who love drawing and are planning to apply to medical school, so we can collectively complete a series of works. Once published, I hope the book will receive guidance from professional medical experts.

All the pictures in this book are hand-drawn by the author, then scanned and labelled with annotations, including simple functional descriptions. To attract a broader audience, this picture book provides bilingual annotations in both Chinese and English. All profits from the book will be donated and dedicated to medical education.

作者简介
About the Author

徐存一，2008 年 12 月出生于英国牛津。在中国北京史家小学完成了小学学习，在中国北京市第二中学分校完成初中学习，曾就读于北京汇文高中，现在英国莱斯特文法学校学习。

Xu Cunyi was born in December 2008 in Oxford, UK. He completed his primary education at Shijia Primary School in Beijing and his junior school education at Beijing No. 2 Middle School Junior Division. He then attended Peking Academy High School in China. Xu is now studying at Leicester Grammar School in the UK.

目　录
Contents

第一章　头部结构

Chapter 1 Structures of Head

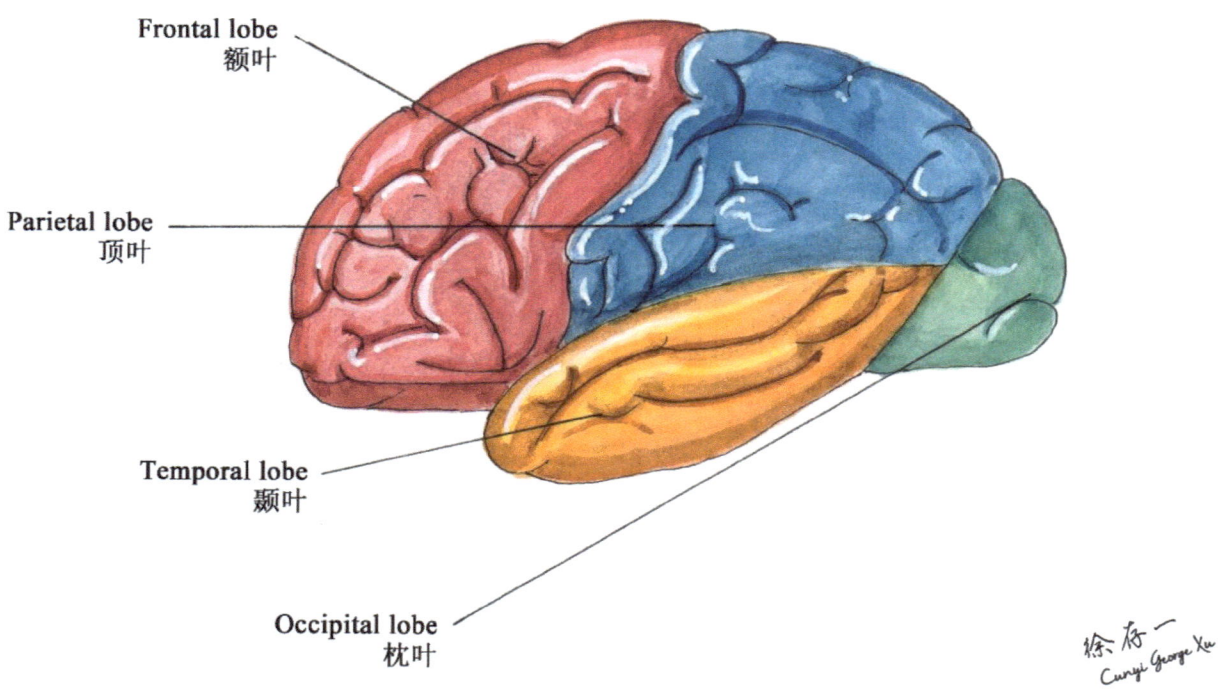

Frontal lobe
额叶

Parietal lobe
顶叶

Temporal lobe
颞叶

Occipital lobe
枕叶

徐存一
Cunyi George Xu

大脑皮层：调节躯体运动和功能的最高级中枢。

The cerebral cortex: the ultimate control centre for body movements and functions.

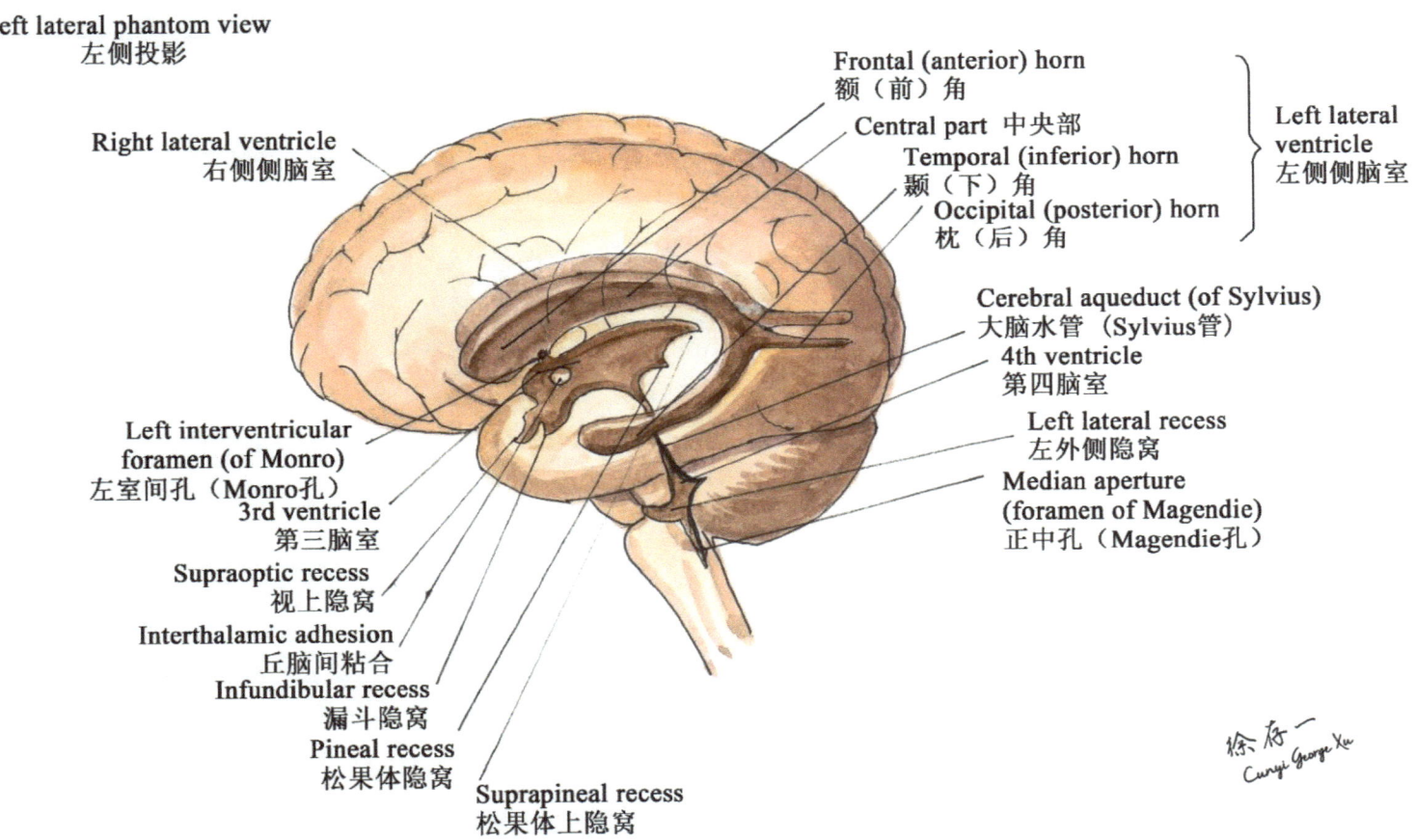

Left lateral phantom view
左侧投影

Right lateral ventricle
右侧侧脑室

Frontal (anterior) horn
额（前）角

Central part 中央部

Temporal (inferior) horn
颞（下）角

Occipital (posterior) horn
枕（后）角

Left lateral
ventricle
左侧侧脑室

Cerebral aqueduct (of Sylvius)
大脑水管（Sylvius管）

4th ventricle
第四脑室

Left lateral recess
左外侧隐窝

Median aperture
(foramen of Magendie)
正中孔（Magendie孔）

Left interventricular
foramen (of Monro)
左室间孔（Monro孔）

3rd ventricle
第三脑室

Supraoptic recess
视上隐窝

Interthalamic adhesion
丘脑间粘合

Infundibular recess
漏斗隐窝

Pineal recess
松果体隐窝

Suprapineal recess
松果体上隐窝

Cunyi George Xu

脑室：主要负责生产和循环脑脊液，为大脑提供营养、废物排除、缓冲保护以及维持颅内压稳定。

Ventricle: production and circulation of cerebrospinal fluid, providing nutrition for the brain, eliminating waste, offering protection, and maintaining stable intracranial pressure.

Choroid plexus of lateral ventricle (phantom) 侧脑室脉络丛（投影）

Superior cerebral veins 大脑上静脉

Dura mater
硬脑膜

Cistern of corpus callosum
胼胝体池

Superior sagittal sinus 上矢状窦

Subarachnoid space 蛛网膜下腔

Arachnoid granulations
蛛网膜粒

Straight sinus 直窦

Quadrigeminal cistern
(with great cerebral vein) (of Galen)
四叠体池（与大脑大静脉）（Galen静脉）

Posterior cerebellomedullary cistern
小脑延髓池

Median aperture (foramen of Magendie)
正中孔（Magendie孔）

Cerebral aqueduct (of Sylvius) 大脑水管（Sylvius孔）

Lateral aperture (foramen of Luschka) 外侧孔（Luschka孔）

Choroid plexus of 4th ventricle 第四脑室脉络丛

Dura mater 硬脑膜

Subarachnoid space 蛛网膜下腔

Central canal of spinal cord
脊髓中央管

脑脊液循环：为大脑提供营养、清除废物、保护大脑免受物理损伤，并维持颅内压力平衡。

Cerebrospinal fluid circulation: supplies nutrients to the brain, removes waste, protects the brain from physical damage, and maintains the balance of intracranial pressure.

Structure of retina: schema
视网膜结构：示意图

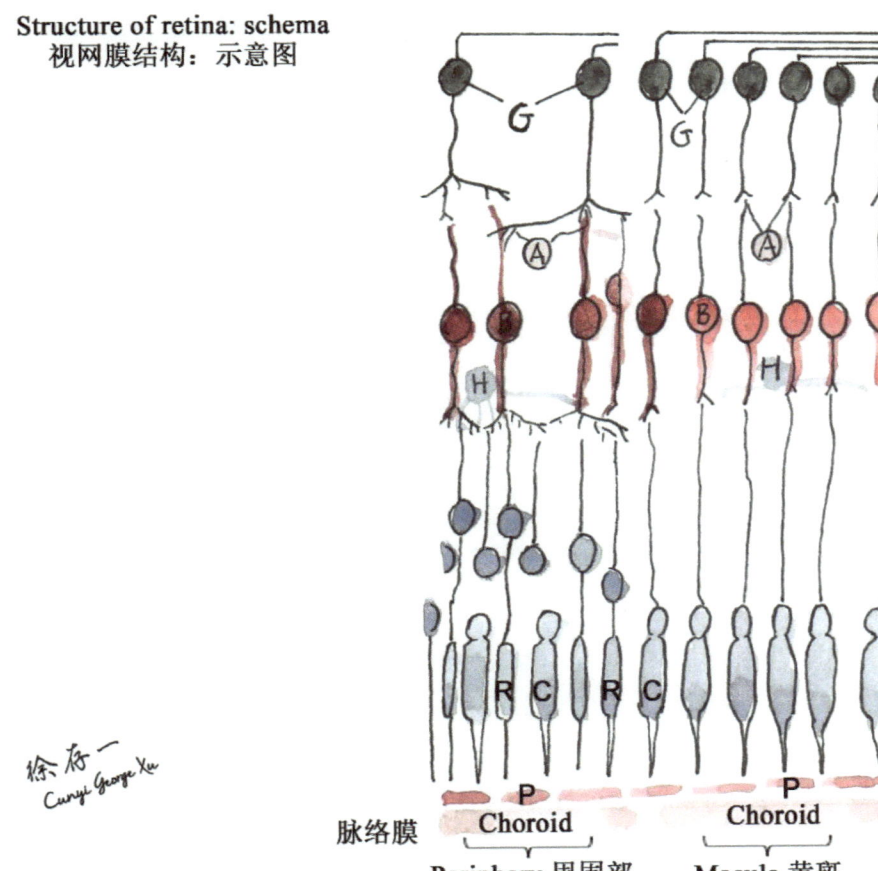

A   Amacrine cells 无长突细胞
B   Bipolar cells 双极细胞
C   Cones 视锥细胞
G   Ganglion cells 节细胞
H   Horizontal cells 水平细胞
P   Pigment cells 色素细胞
R   Rods 视杆细胞

视网膜：负责光信号的接收和初步处理。

Retina: responsible for receiving light signals and initial processing.

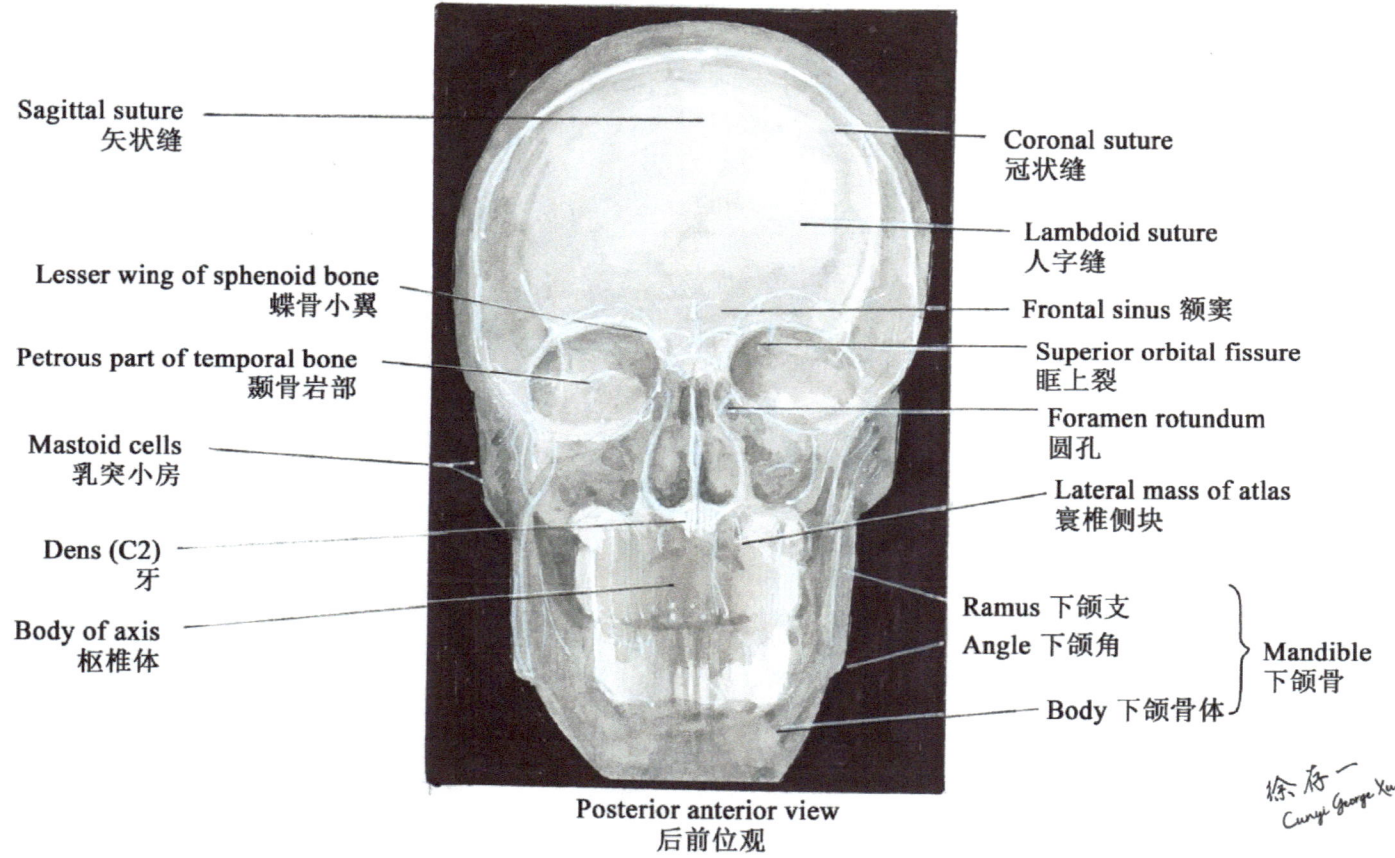

Sagittal suture
矢状缝

Coronal suture
冠状缝

Lambdoid suture
人字缝

Lesser wing of sphenoid bone
蝶骨小翼

Frontal sinus 额窦

Petrous part of temporal bone
颞骨岩部

Superior orbital fissure
眶上裂

Foramen rotundum
圆孔

Mastoid cells
乳突小房

Lateral mass of atlas
寰椎侧块

Dens (C2)
牙

Body of axis
枢椎体

Ramus 下颌支
Angle 下颌角

Mandible
下颌骨

Body 下颌骨体

Posterior anterior view
后前位观

颅骨：保护大脑和感觉器官，为头部的肌肉和血管提供支撑结构。

Skull: protects the brain and sensory organs, and provides a supportive structure for the muscles and blood vessels in the head.

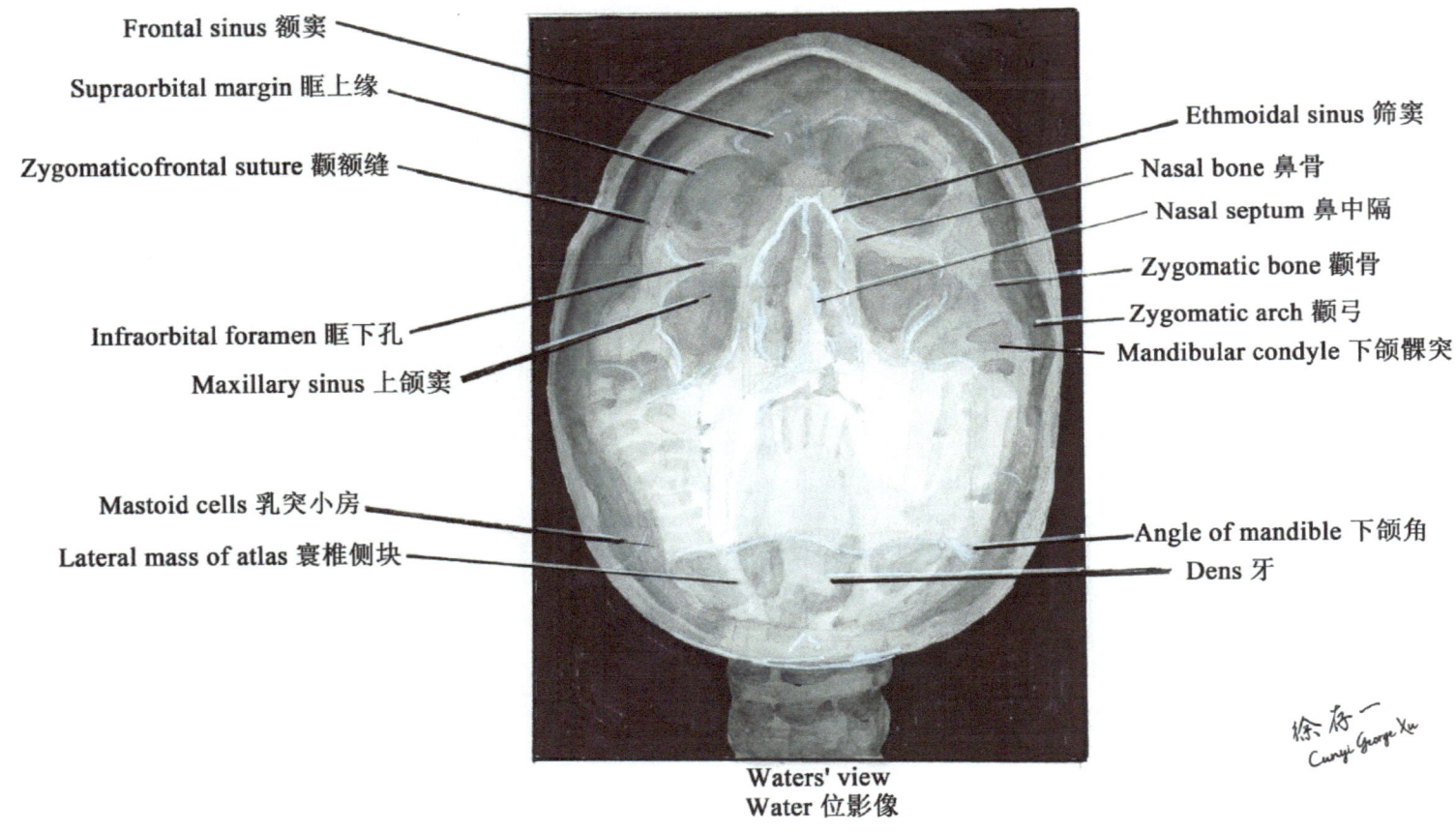

Frontal sinus 额窦

Supraorbital margin 眶上缘

Zygomaticofrontal suture 颧额缝

Infraorbital foramen 眶下孔

Maxillary sinus 上颌窦

Mastoid cells 乳突小房

Lateral mass of atlas 寰椎侧块

Ethmoidal sinus 筛窦

Nasal bone 鼻骨

Nasal septum 鼻中隔

Zygomatic bone 颧骨

Zygomatic arch 颧弓

Mandibular condyle 下颌髁突

Angle of mandible 下颌角

Dens 牙

徐存一
Cunyi George Xu

Waters' view
Water 位影像

颅骨：保护大脑和感觉器官，为头部的肌肉和血管提供支撑结构。

Skull: protects the brain and sensory organs, and provides a supportive structure for the muscles and blood vessels in the head.

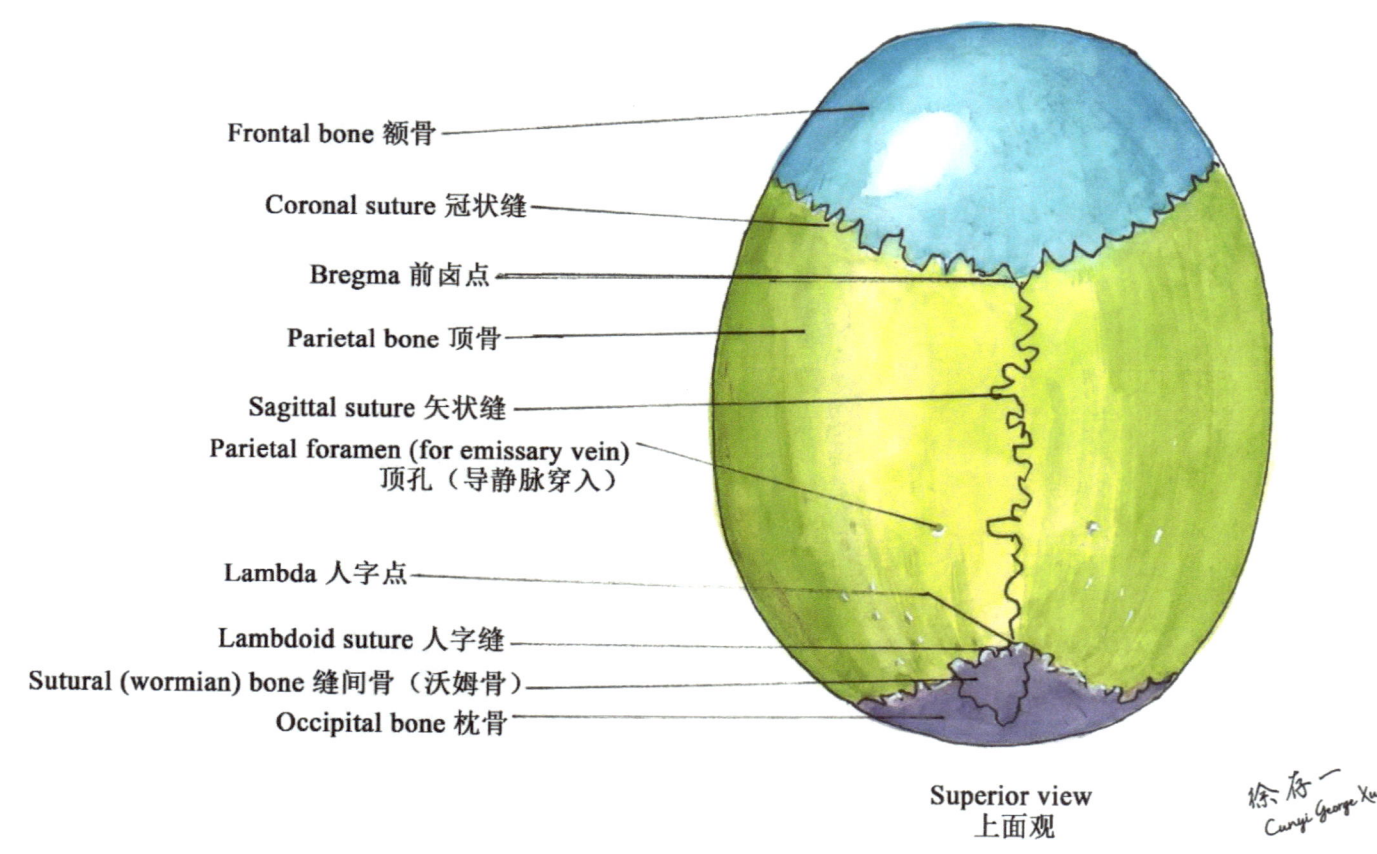

Frontal bone 额骨

Coronal suture 冠状缝

Bregma 前卤点

Parietal bone 顶骨

Sagittal suture 矢状缝

Parietal foramen (for emissary vein)
顶孔（导静脉穿入）

Lambda 人字点

Lambdoid suture 人字缝

Sutural (wormian) bone 缝间骨（沃姆骨）

Occipital bone 枕骨

Superior view
上面观

颅盖：保护大脑免受外界伤害，为大脑提供稳定的物理环境。

Skullcap: protects the brain from external injuries and provides a stable physical environment for the brain.

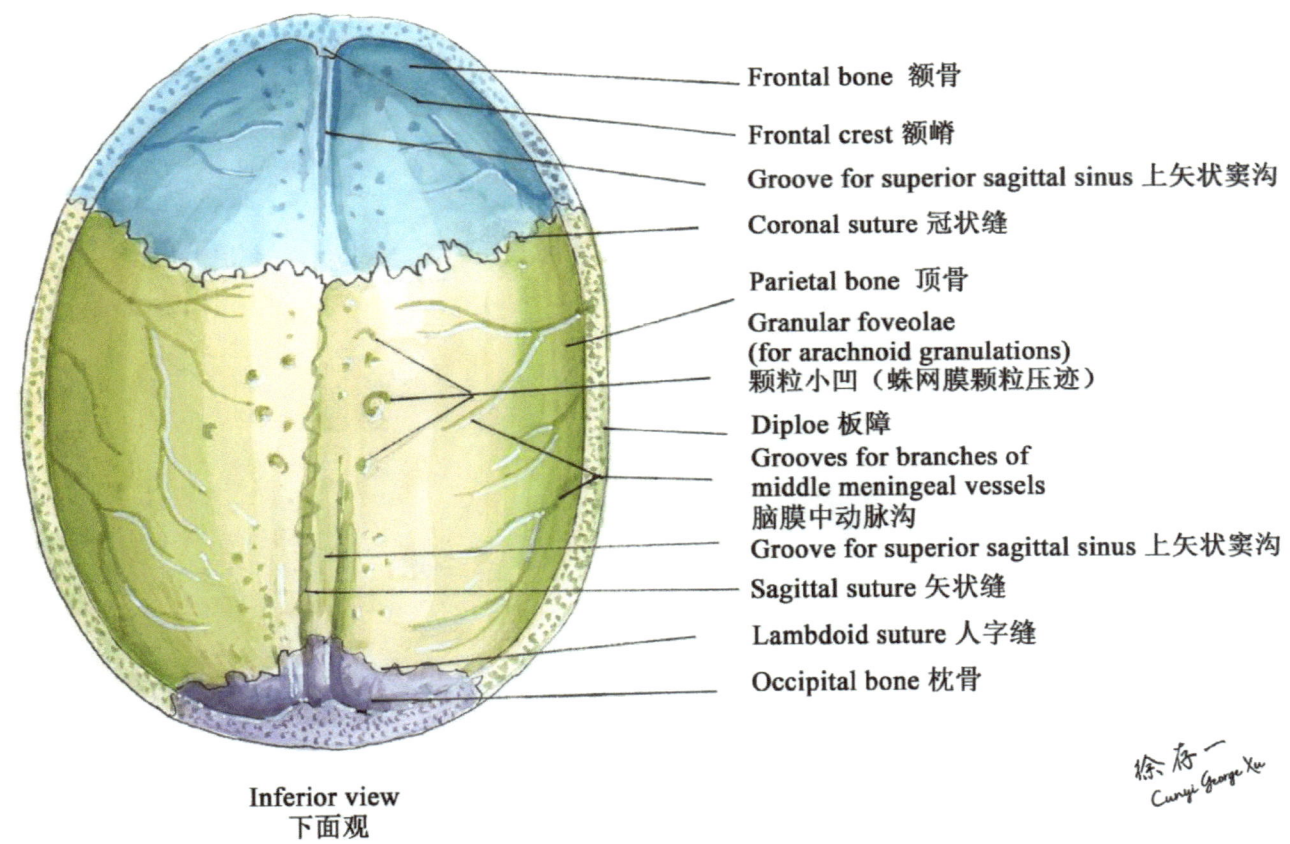

Frontal bone 额骨

Frontal crest 额嵴

Groove for superior sagittal sinus 上矢状窦沟

Coronal suture 冠状缝

Parietal bone 顶骨

Granular foveolae
(for arachnoid granulations)
颗粒小凹（蛛网膜颗粒压迹）

Diploe 板障

Grooves for branches of
middle meningeal vessels
脑膜中动脉沟

Groove for superior sagittal sinus 上矢状窦沟

Sagittal suture 矢状缝

Lambdoid suture 人字缝

Occipital bone 枕骨

Inferior view
下面观

颅盖：保护大脑免受外界伤害，为大脑提供稳定的物理环境。

Skullcap: protects the brain from external injuries and provides a stable physical environment for the brain.

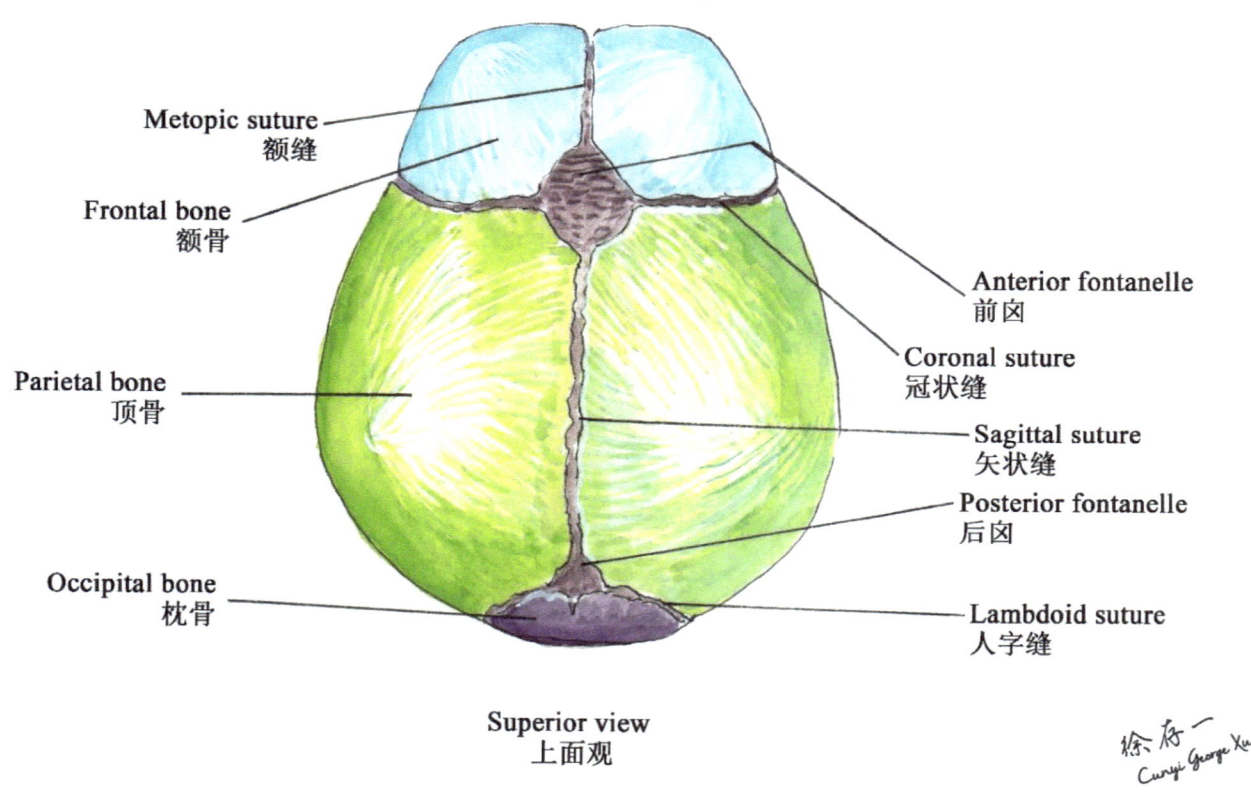

Metopic suture
额缝

Frontal bone
额骨

Anterior fontanelle
前囟

Coronal suture
冠状缝

Parietal bone
顶骨

Sagittal suture
矢状缝

Posterior fontanelle
后囟

Occipital bone
枕骨

Lambdoid suture
人字缝

Superior view
上面观

新生儿颅：在出生时允许头骨在通过产道时变形，并在大脑发育过程中为其提供空间。

Neonatal skull: allows the skull to deform during birth and provides space for brain development.

第二章　头部血管和神经

Chapter 2 Blood Vessels and Nerves of Head

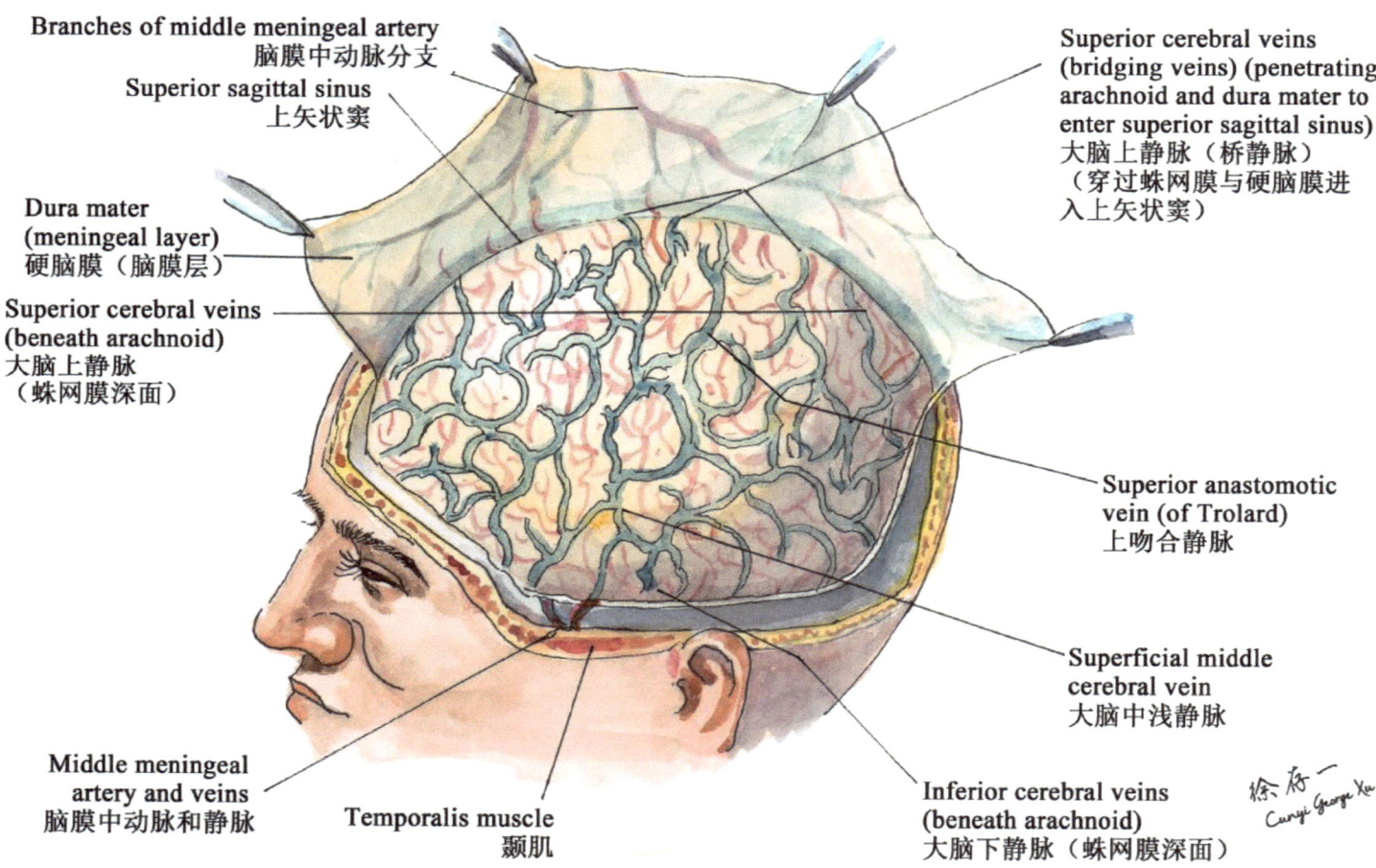

Branches of middle meningeal artery
脑膜中动脉分支

Superior sagittal sinus
上矢状窦

Dura mater
(meningeal layer)
硬脑膜（脑膜层）

Superior cerebral veins
(beneath arachnoid)
大脑上静脉
（蛛网膜深面）

Superior cerebral veins
(bridging veins) (penetrating
arachnoid and dura mater to
enter superior sagittal sinus)
大脑上静脉（桥静脉）
（穿过蛛网膜与硬脑膜进
入上矢状窦）

Superior anastomotic
vein (of Trolard)
上吻合静脉

Superficial middle
cerebral vein
大脑中浅静脉

Middle meningeal
artery and veins
脑膜中动脉和静脉

Temporalis muscle
颞肌

Inferior cerebral veins
(beneath arachnoid)
大脑下静脉（蛛网膜深面）

脑膜与大脑上静脉：为大脑提供血液供应和排出静脉血，并保护大脑免受外部损伤。

Meninges and superior cerebral veins: provide blood supply and drain venous blood from the brain, as well as protect the brain from external injuries.

Magnetic resonance venography (MRV): schema
磁共振静脉造影成像：示意图

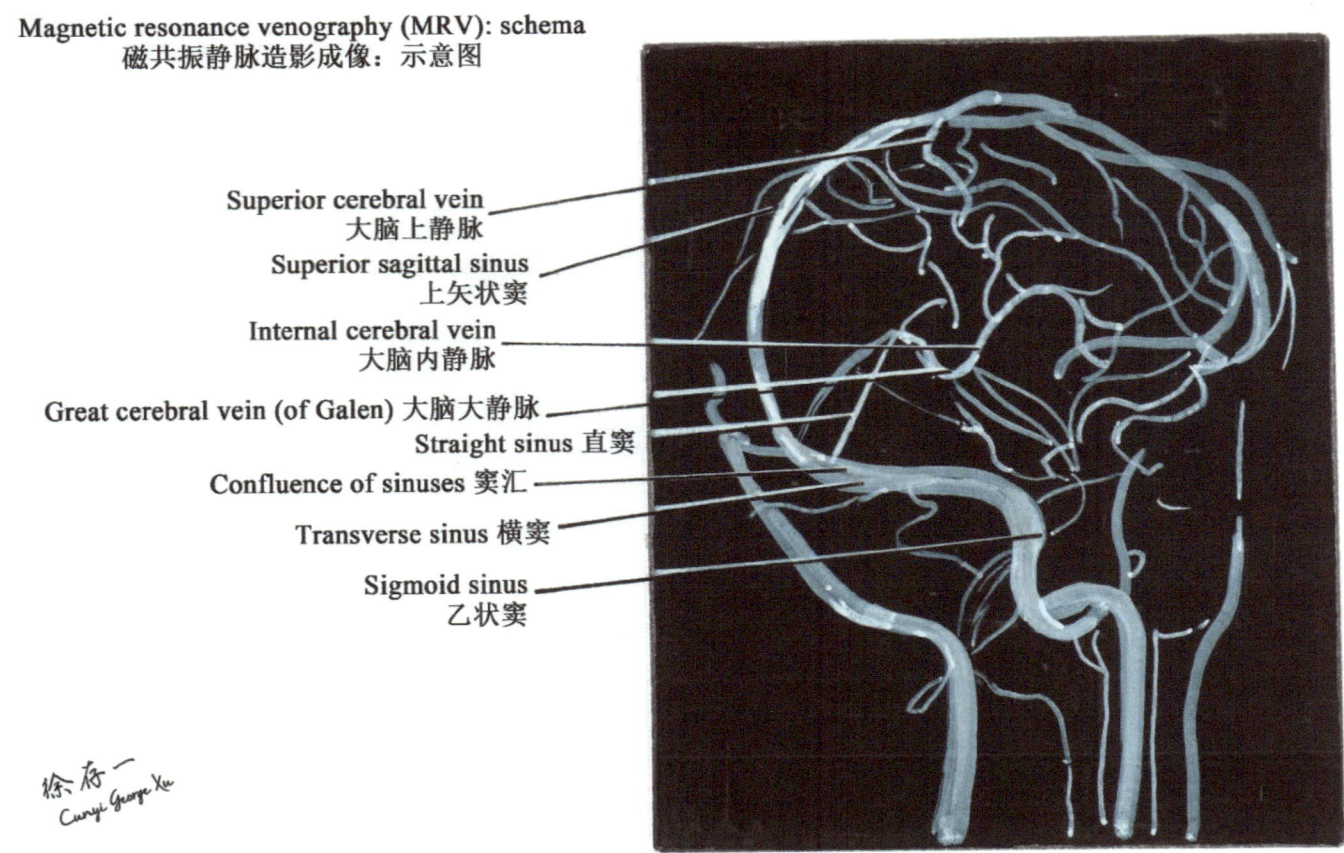

Superior cerebral vein
大脑上静脉

Superior sagittal sinus
上矢状窦

Internal cerebral vein
大脑内静脉

Great cerebral vein (of Galen) 大脑大静脉

Straight sinus 直窦

Confluence of sinuses 窦汇

Transverse sinus 横窦

Sigmoid sinus
乙状窦

徐存一
Cunyi George Xu

大脑静脉系统：将脑内的静脉血液排出，确保大脑的血液循环正常运作。

Cerebral venous system: expels venous blood from the brain to ensure normal blood circulation in the brain.

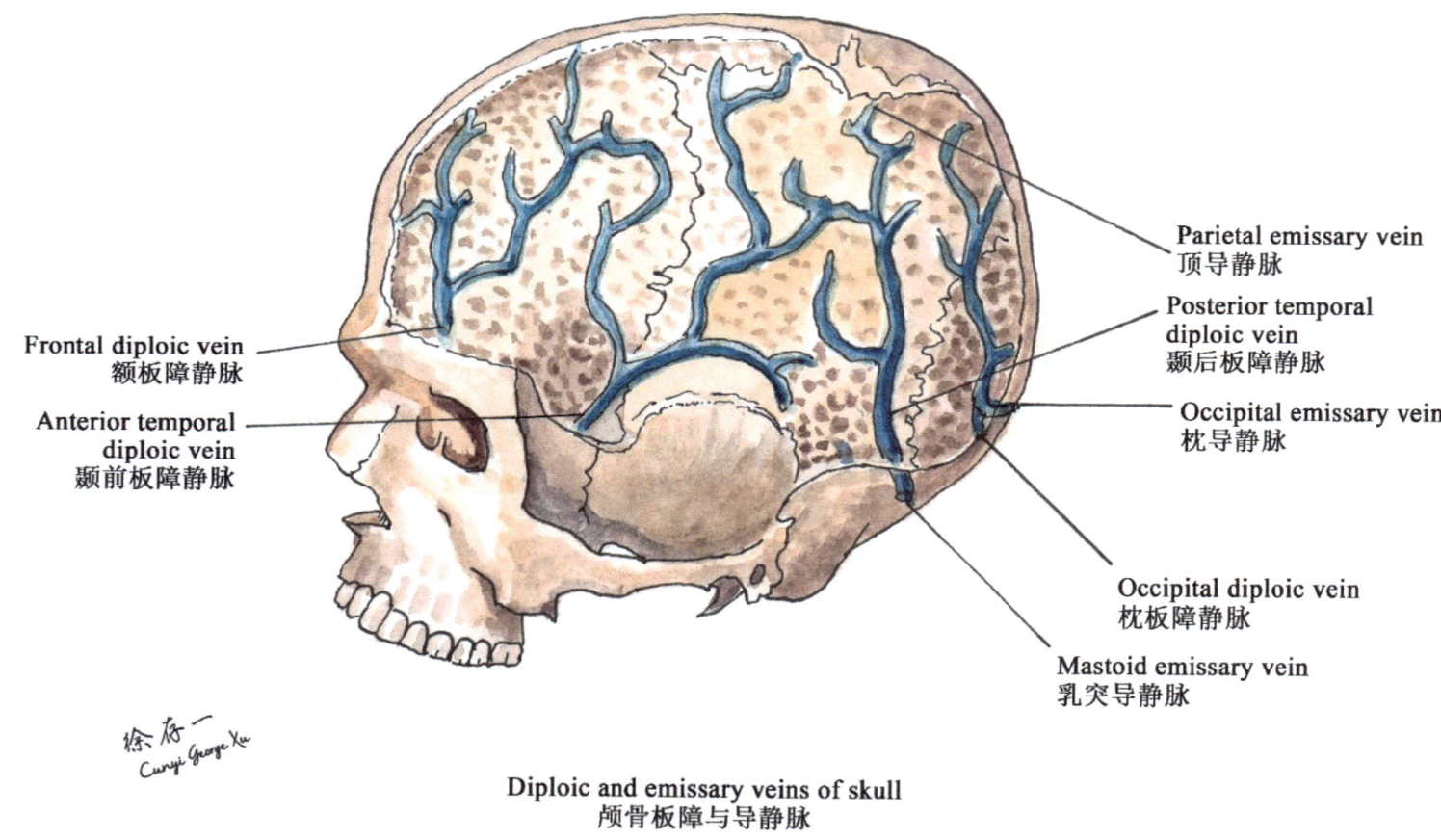

Parietal emissary vein
顶导静脉

Posterior temporal
diploic vein
颞后板障静脉

Occipital emissary vein
枕导静脉

Frontal diploic vein
额板障静脉

Anterior temporal
diploic vein
颞前板障静脉

Occipital diploic vein
枕板障静脉

Mastoid emissary vein
乳突导静脉

徐存一
Cunyi George Xu

Diploic and emissary veins of skull
颅骨板障与导静脉

颅骨板障与导静脉： 帮助颅骨内外部的静脉血循环，减轻颅内压。

Diploic and emissary veins of the skull: assist in venous blood circulation between the interior and exterior of the skull and reduce intracranial pressure.

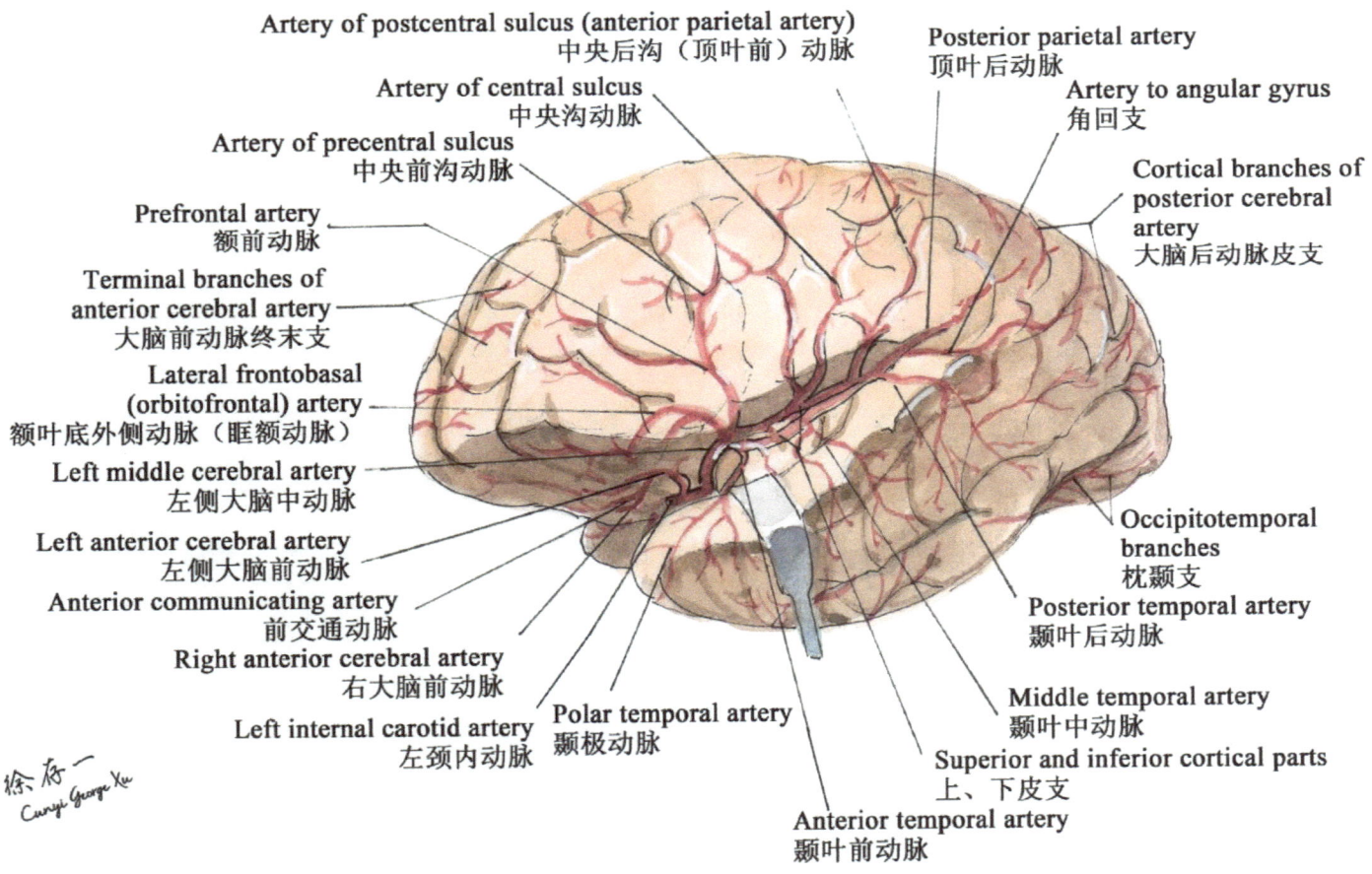

Artery of postcentral sulcus (anterior parietal artery)
中央后沟（顶叶前）动脉

Artery of central sulcus
中央沟动脉

Artery of precentral sulcus
中央前沟动脉

Prefrontal artery
额前动脉

Terminal branches of
anterior cerebral artery
大脑前动脉终末支

Lateral frontobasal
(orbitofrontal) artery
额叶底外侧动脉（眶额动脉）

Left middle cerebral artery
左侧大脑中动脉

Left anterior cerebral artery
左侧大脑前动脉

Anterior communicating artery
前交通动脉

Right anterior cerebral artery
右大脑前动脉

Left internal carotid artery
左颈内动脉

Polar temporal artery
颞极动脉

Anterior temporal artery
颞叶前动脉

Posterior parietal artery
顶叶后动脉

Artery to angular gyrus
角回支

Cortical branches of
posterior cerebral
artery
大脑后动脉皮支

Occipitotemporal
branches
枕颞支

Posterior temporal artery
颞叶后动脉

Middle temporal artery
颞叶中动脉

Superior and inferior cortical parts
上、下皮支

Cunyi George Xu

大脑动脉：负责将氧气和营养物质输送到大脑，清除代谢废物，以维持大脑功能。

Cerebral artery: delivers oxygen and nutrients to the brain and clears metabolic waste to support brain function.

Long medial striate artery (recurrent artery of Heubner)
纹状体长内侧动脉（Heubner返动脉）

Anteromedial central (perforating) arteries
前内侧中央（穿）动脉

Anterolateral central (lenticulostriate) arteries
前外侧中央（豆纹）动脉

Superior hypophyseal artery
垂体上动脉

Inferior hypophyseal artery
垂体下动脉

Anterior choroidal artery
脉络丛前动脉

Posteromedial central (perforating) arteries
后内侧中央（穿）动脉

Thalamoperforating artery
丘脑穿动脉

Posteromedial central (paramedian) arteries
后内侧中央（旁正中）动脉

Labyrinthine artery
迷路动脉

Anterior cerebral artery (A2 segment)
大脑前动脉（A2段）

Anterior communicating artery
前交通动脉

Ophthalmic artery
眼动脉

Internal carotid artery 颈内动脉

Middle cerebral artery
大脑中动脉

Posterior communicating artery 后交通动脉

Posterior cerebral artery 大脑后动脉
(P2 segment) （P2段）
(P1 segment) （P1段）

Superior cerebellar artery
小脑上动脉

Basilar artery 基底动脉

Pontine arteries 脑桥动脉

Anterior inferior cerebellar artery
小脑下前动脉

Vertebral artery 椎动脉

Vessels dissected out: inferior view
分离出的血管：下面观

徐存一
Cunyi George Xu

大脑动脉：为大脑提供血液供应。

Cerebral artery: provides blood supply to the brain.

ATLAS OF HUMAN ANATOMY
FOR JUNIOR PREMEDICAL STUDENTS

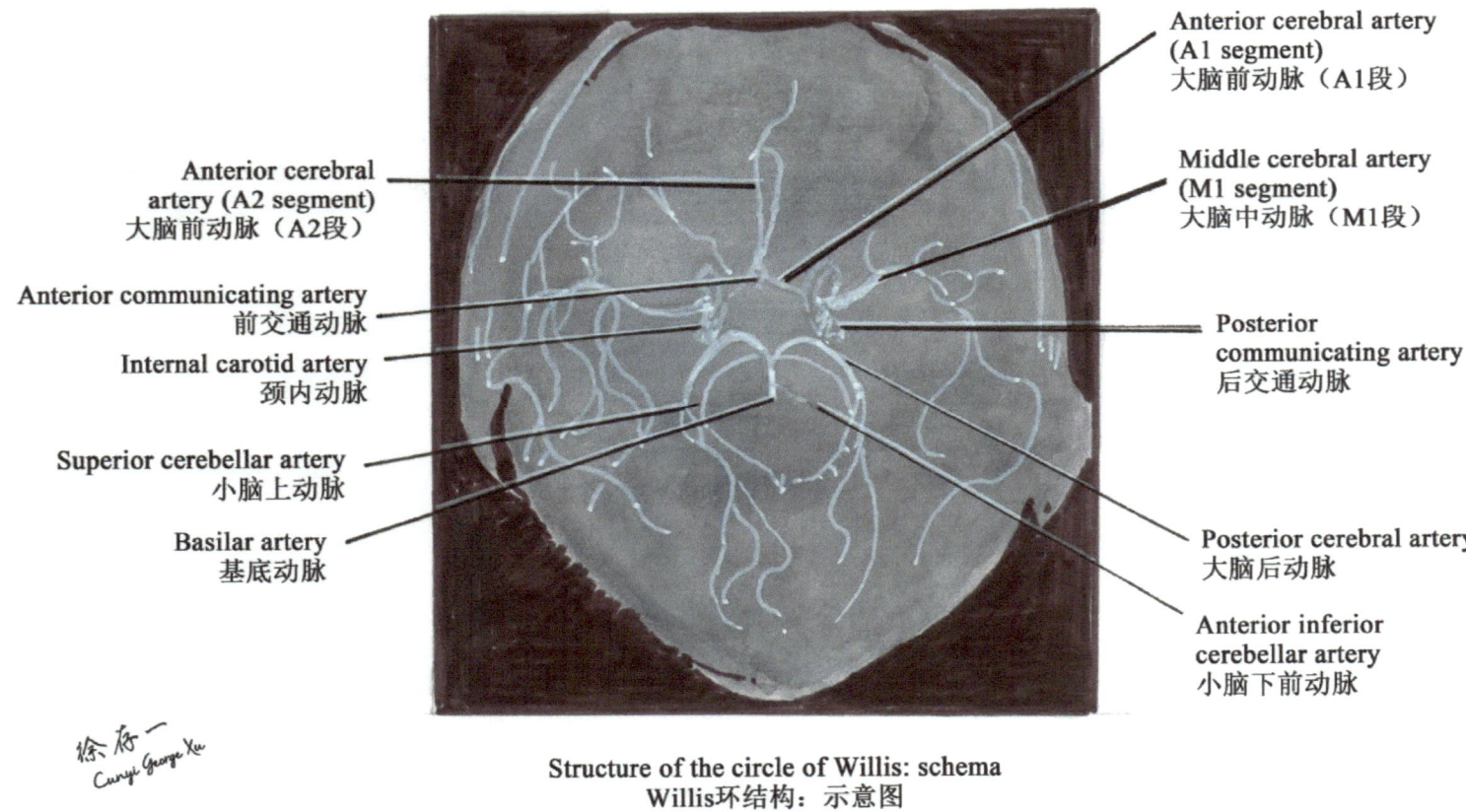

Anterior cerebral artery
(A1 segment)
大脑前动脉（A1段）

Middle cerebral artery
(M1 segment)
大脑中动脉（M1段）

Anterior cerebral
artery (A2 segment)
大脑前动脉（A2段）

Anterior communicating artery
前交通动脉

Internal carotid artery
颈内动脉

Posterior
communicating artery
后交通动脉

Superior cerebellar artery
小脑上动脉

Basilar artery
基底动脉

Posterior cerebral artery
大脑后动脉

Anterior inferior
cerebellar artery
小脑下前动脉

Structure of the circle of Willis: schema
Willis环结构：示意图

Willis 环： 确保脑部各区域在血流受阻时仍能获得足够的血液供应。

Circle of Willis: ensures sufficient blood supply to various regions of the brain, even when blood flow is obstructed.

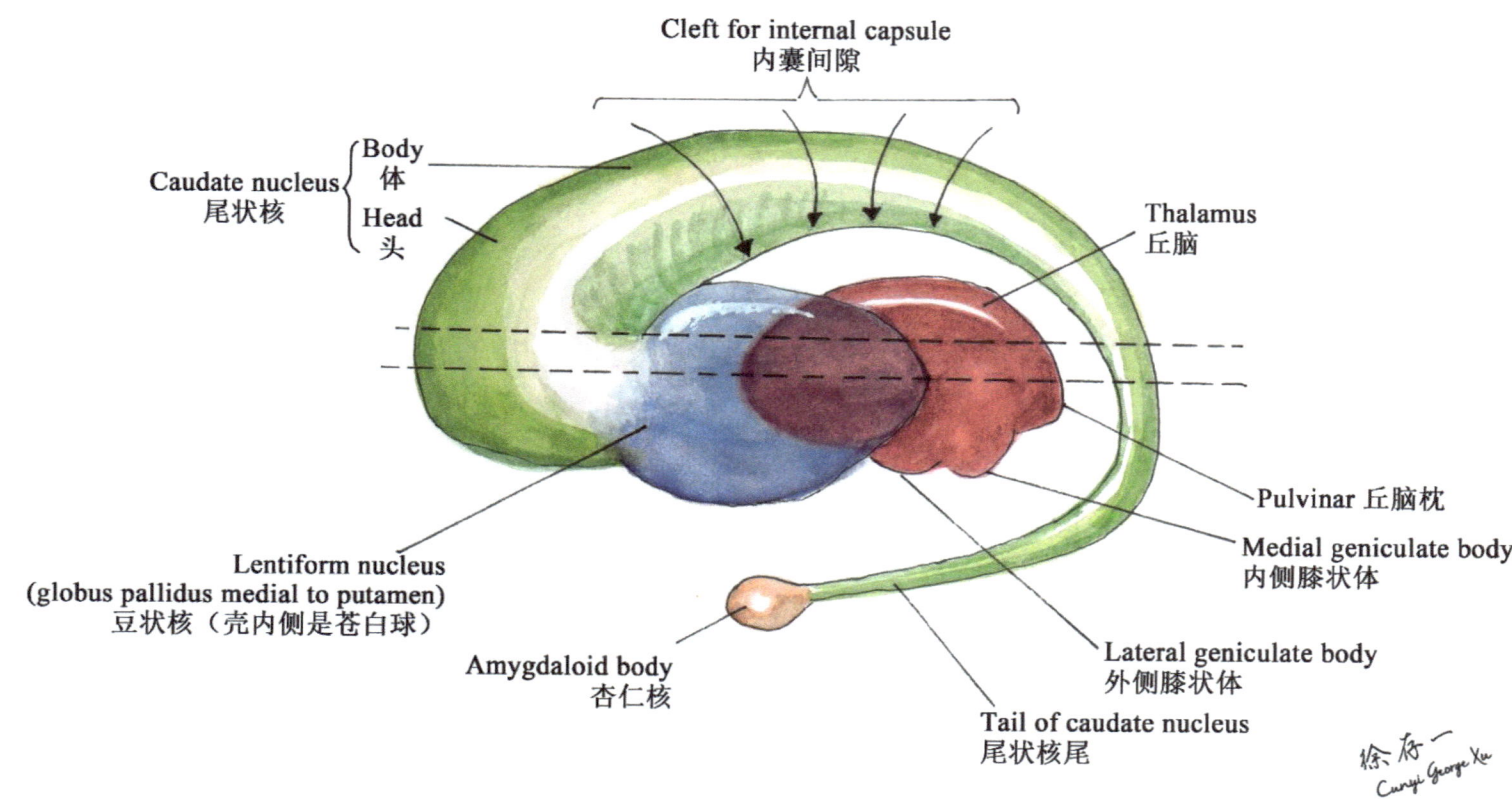

Cleft for internal capsule
内囊间隙

Caudate nucleus
尾状核 { Body 体 / Head 头 }

Thalamus
丘脑

Lentiform nucleus
(globus pallidus medial to putamen)
豆状核（壳内侧是苍白球）

Amygdaloid body
杏仁核

Pulvinar 丘脑枕

Medial geniculate body
内侧膝状体

Lateral geniculate body
外侧膝状体

Tail of caudate nucleus
尾状核尾

Interrelationship of thalamus, lentiform nucleus, caudate nucleus, and amygdaloid body (schema): left lateral view
丘脑、豆状核、尾状核与杏仁核的位置关系：左侧面观

基底神经节：控制自主运动，参与记忆、情感和奖励学习等高级认知功能。

Basal ganglia: control autonomous movement and participate in advanced cognitive functions such as memory, emotion, and reward learning.

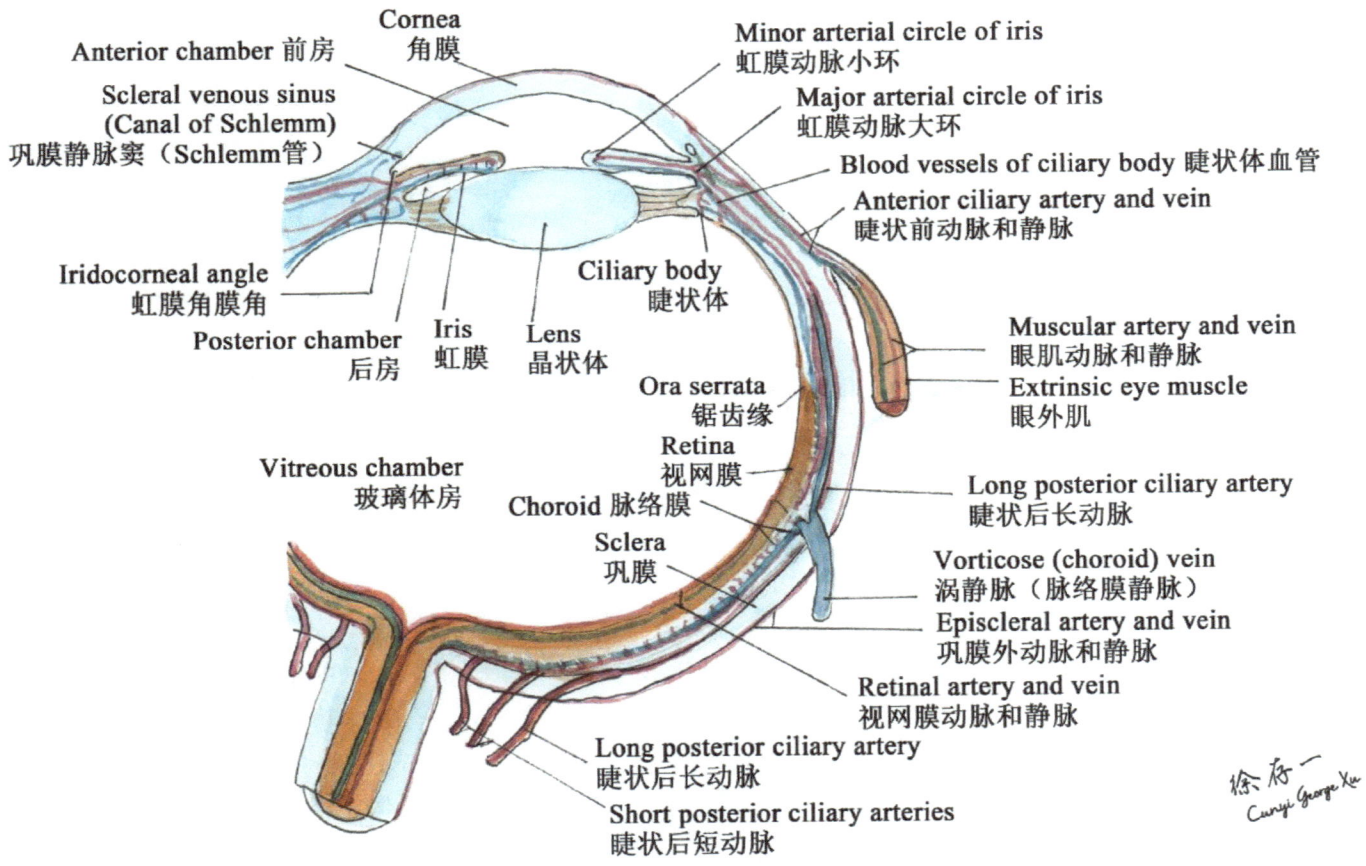

Cornea
角膜

Anterior chamber 前房

Minor arterial circle of iris
虹膜动脉小环

Scleral venous sinus
(Canal of Schlemm)
巩膜静脉窦（Schlemm管）

Major arterial circle of iris
虹膜动脉大环

Blood vessels of ciliary body 睫状体血管

Anterior ciliary artery and vein
睫状前动脉和静脉

Iridocorneal angle
虹膜角膜角

Ciliary body
睫状体

Muscular artery and vein
眼肌动脉和静脉

Extrinsic eye muscle
眼外肌

Posterior chamber
后房

Iris
虹膜

Lens
晶状体

Ora serrata
锯齿缘

Retina
视网膜

Long posterior ciliary artery
睫状后长动脉

Vitreous chamber
玻璃体房

Choroid 脉络膜

Sclera
巩膜

Vorticose (choroid) vein
涡静脉（脉络膜静脉）

Episcleral artery and vein
巩膜外动脉和静脉

Retinal artery and vein
视网膜动脉和静脉

Long posterior ciliary artery
睫状后长动脉

Short posterior ciliary arteries
睫状后短动脉

Cunyi George Xu

眼内的动脉和静脉：为眼睛提供氧气和营养，清除废物和二氧化碳，维持眼睛的健康和功能。

Intraocular arteries and veins: provide oxygen and nutrients to the eyes, remove waste and carbon dioxide, and maintain eye health and function.

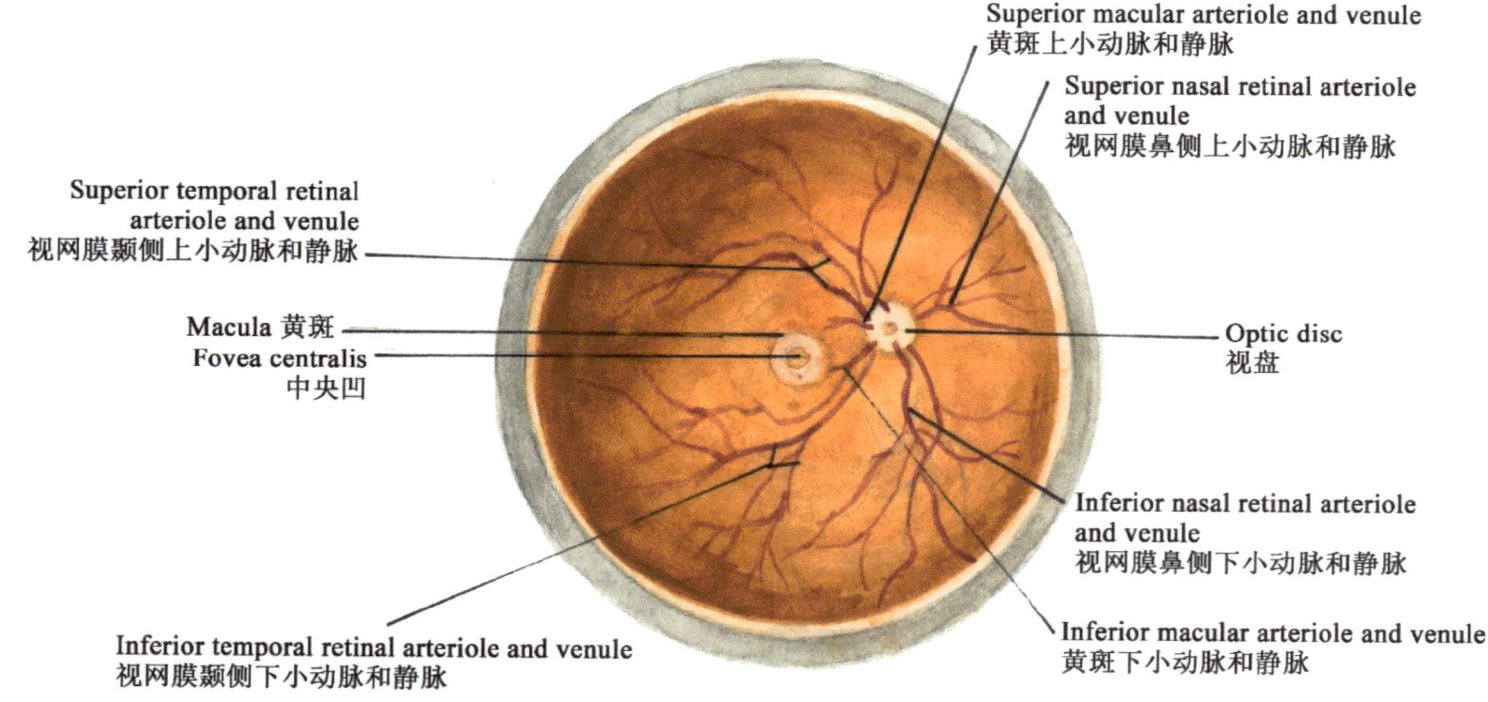

Superior macular arteriole and venule
黄斑上小动脉和静脉

Superior nasal retinal arteriole and venule
视网膜鼻侧上小动脉和静脉

Superior temporal retinal arteriole and venule
视网膜颞侧上小动脉和静脉

Macula 黄斑
Fovea centralis
中央凹

Optic disc
视盘

Inferior nasal retinal arteriole and venule
视网膜鼻侧下小动脉和静脉

Inferior temporal retinal arteriole and venule
视网膜颞侧下小动脉和静脉

Inferior macular arteriole and venule
黄斑下小动脉和静脉

Right retinal vessels: funduscopic view
右侧视网膜血管：眼底镜下观

Cunyi George Xu

视网膜血管：为视网膜提供血液供应，保证视觉功能的正常运行。

Retinal vessels: provide blood supply to the retina to ensure the normal functioning of vision.

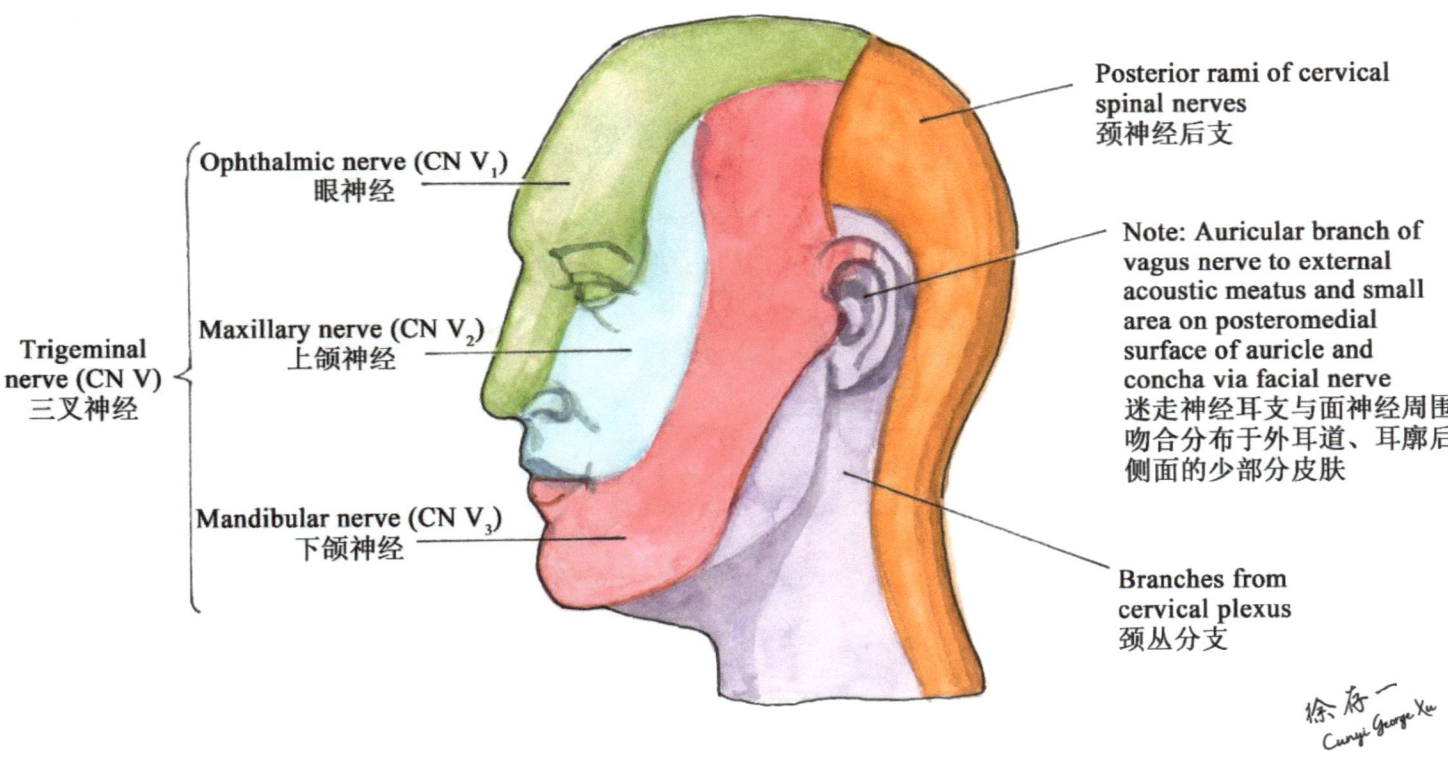

Ophthalmic nerve (CN V₁)
眼神经

Maxillary nerve (CN V₂)
上颌神经

Trigeminal
nerve (CN V)
三叉神经

Mandibular nerve (CN V₃)
下颌神经

Posterior rami of cervical
spinal nerves
颈神经后支

Note: Auricular branch of
vagus nerve to external
acoustic meatus and small
area on posteromedial
surface of auricle and
concha via facial nerve
迷走神经耳支与面神经周围
吻合分布于外耳道、耳廓后
侧面的少部分皮肤

Branches from
cervical plexus
颈丛分支

三叉神经：面部的感觉传导。

Trigeminal nerves: transmit facial sensation.

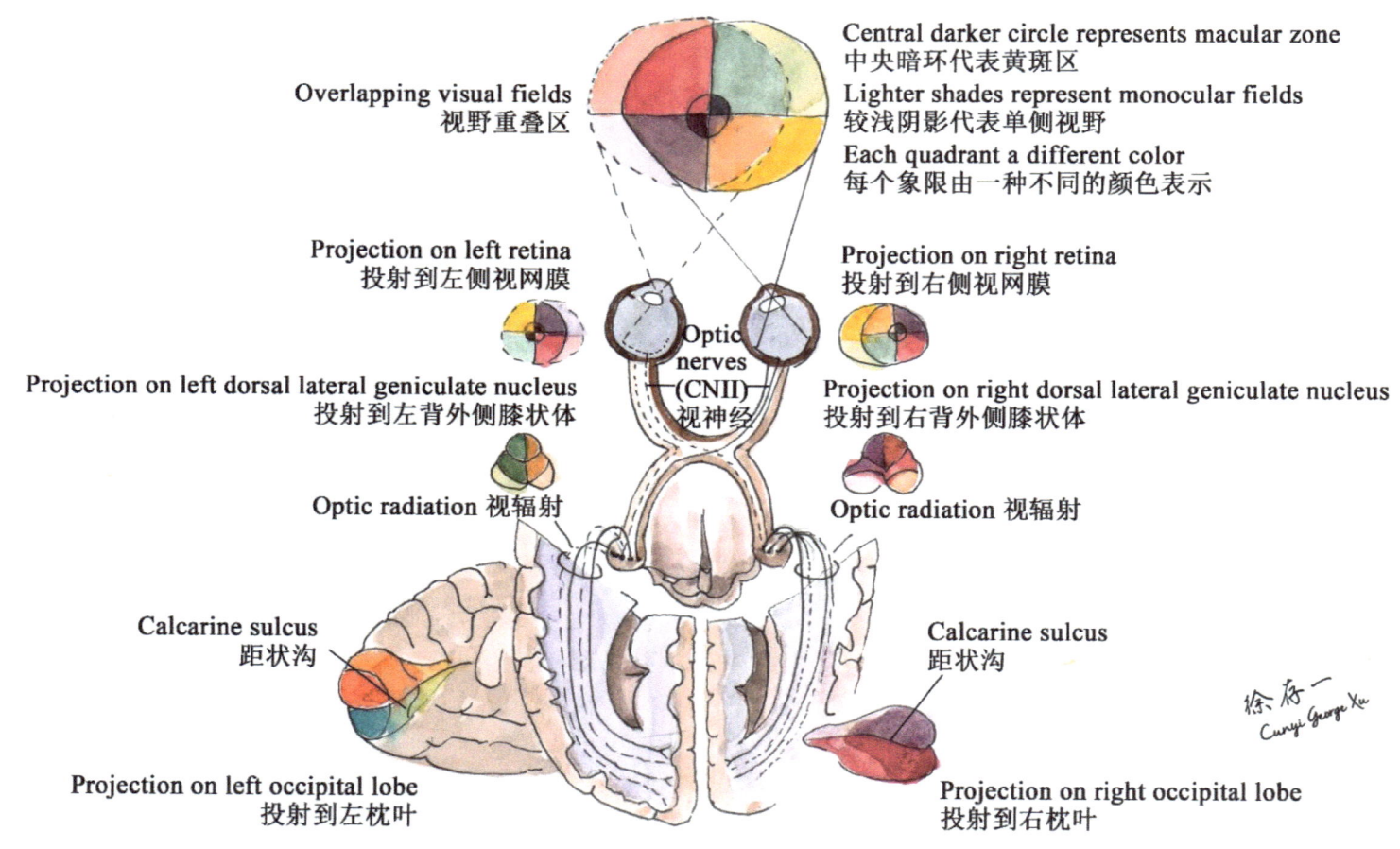

Central darker circle represents macular zone
中央暗环代表黄斑区
Lighter shades represent monocular fields
较浅阴影代表单侧视野
Each quadrant a different color
每个象限由一种不同的颜色表示

Overlapping visual fields
视野重叠区

Projection on left retina
投射到左侧视网膜

Projection on right retina
投射到右侧视网膜

Optic nerves (CNII)
视神经

Projection on left dorsal lateral geniculate nucleus
投射到左背外侧膝状体

Projection on right dorsal lateral geniculate nucleus
投射到右背外侧膝状体

Optic radiation 视辐射

Optic radiation 视辐射

Calcarine sulcus
距状沟

Calcarine sulcus
距状沟

Projection on left occipital lobe
投射到左枕叶

Projection on right occipital lobe
投射到右枕叶

徐存一
Cunyi George Xu

视神经：负责将视网膜的光感受器产生的信号传递到大脑，从而产生视觉。

Optic nerves: transmit the signals generated by the photoreceptors in the retina to the brain, thereby producing vision.

第三章　胸、背和脊髓

Chapter 3 Chest, Back and Spinal Cord

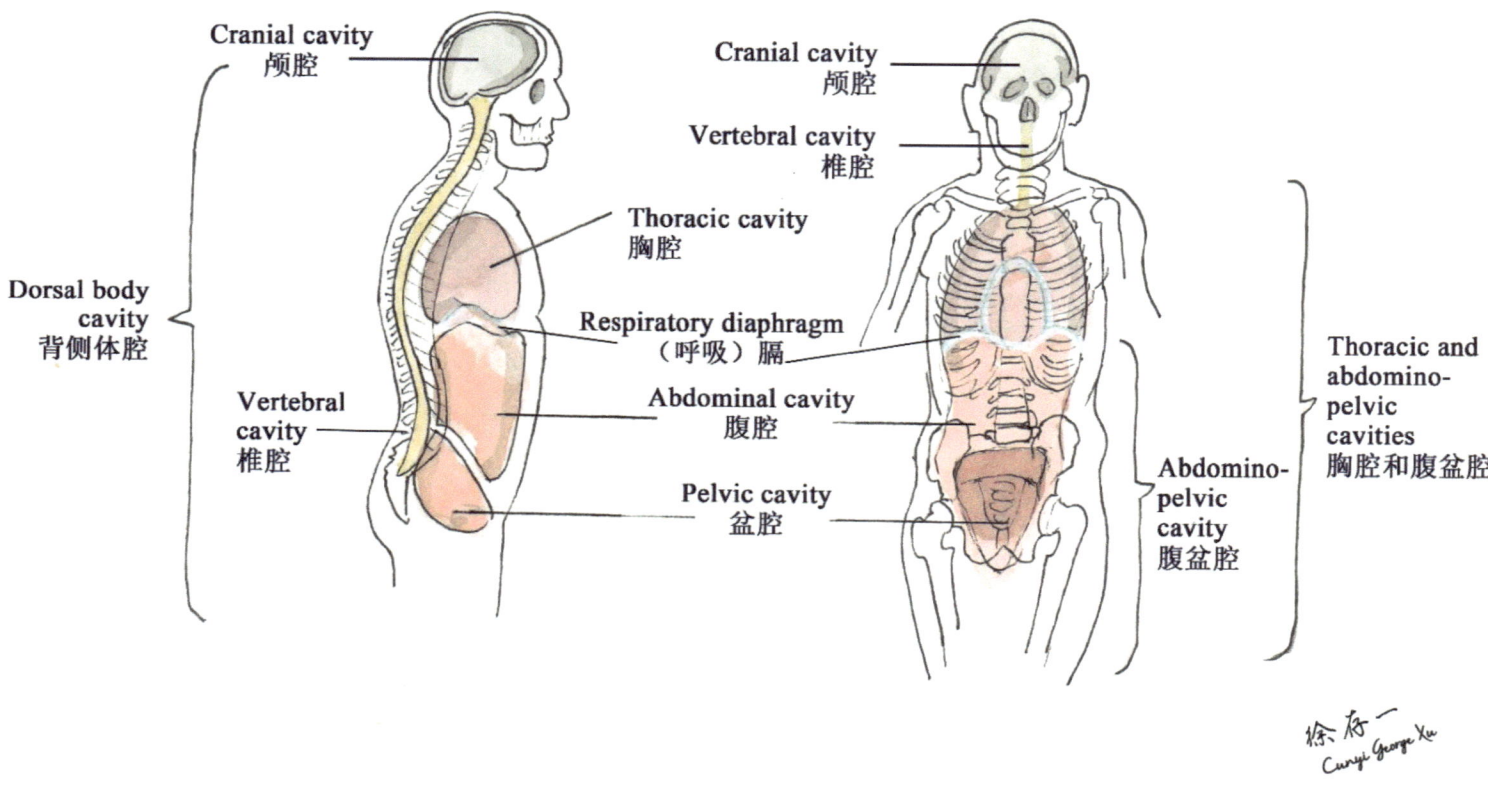

Cranial cavity
颅腔

Dorsal body
cavity
背侧体腔

Vertebral
cavity
椎腔

Cranial cavity
颅腔

Vertebral cavity
椎腔

Thoracic cavity
胸腔

Respiratory diaphragm
（呼吸）膈

Abdominal cavity
腹腔

Pelvic cavity
盆腔

Thoracic and
abdomino-
pelvic
cavities
胸腔和腹盆腔

Abdomino-
pelvic
cavity
腹盆腔

体腔：体内脏器周围的腔隙，增进机体灵活性，保护内部器官。

Cavity: the space around the internal organs that enhances flexibility and protects them.

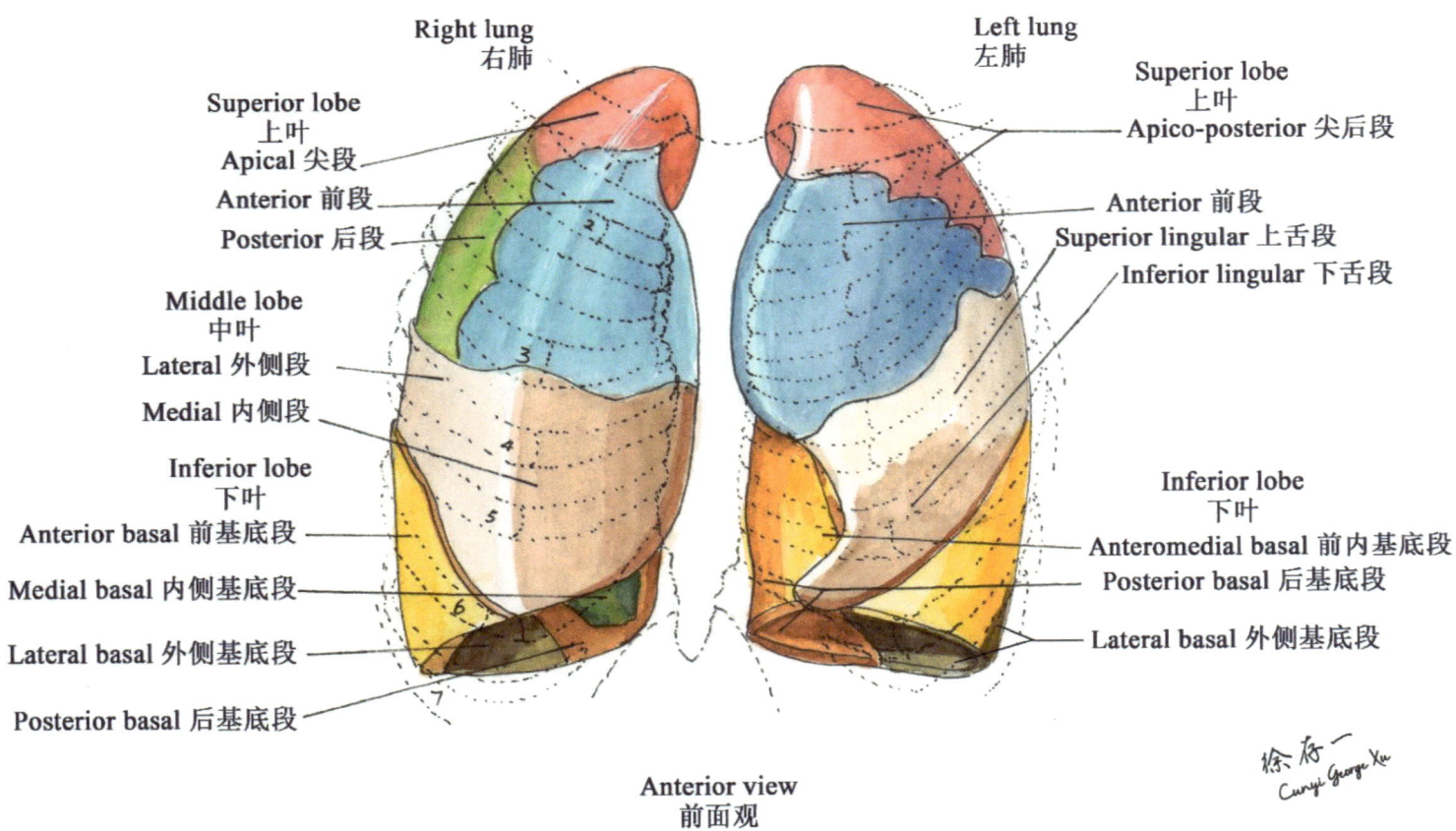

Right lung
右肺

Superior lobe
上叶
Apical 尖段
Anterior 前段
Posterior 后段

Middle lobe
中叶
Lateral 外侧段
Medial 内侧段

Inferior lobe
下叶
Anterior basal 前基底段
Medial basal 内侧基底段
Lateral basal 外侧基底段
Posterior basal 后基底段

Left lung
左肺

Superior lobe
上叶
Apico-posterior 尖后段

Anterior 前段
Superior lingular 上舌段
Inferior lingular 下舌段

Inferior lobe
下叶
Anteromedial basal 前内基底段
Posterior basal 后基底段

Lateral basal 外侧基底段

Anterior view
前面观

肺：人体的呼吸器官，位于胸腔。

Lung: the respiratory organ of the human body, located in the chest

ATLAS OF HUMAN ANATOMY
FOR JUNIOR PREMEDICAL STUDENTS

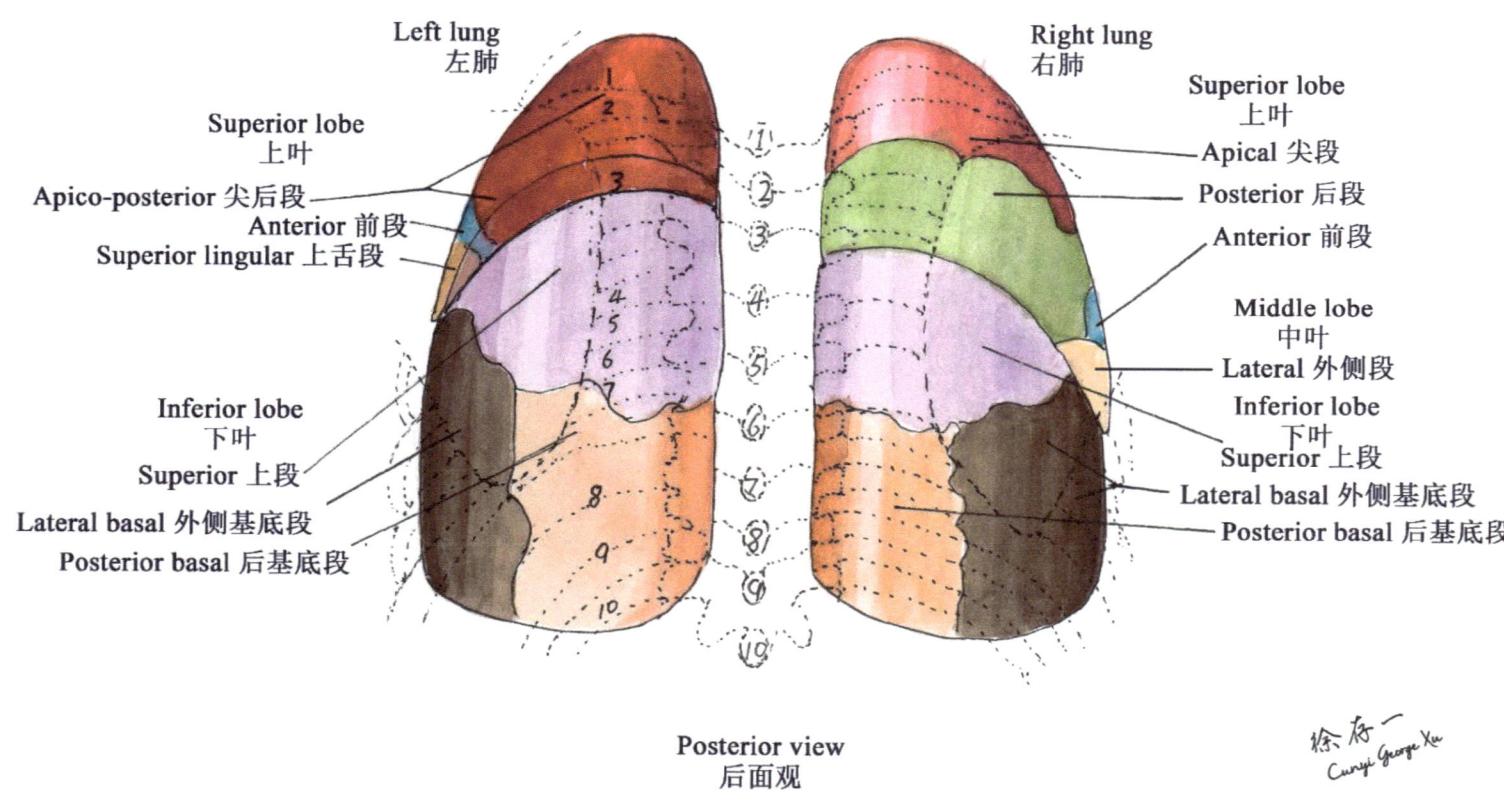

Left lung
左肺

Superior lobe
上叶

Apico-posterior 尖后段
Anterior 前段
Superior lingular 上舌段

Inferior lobe
下叶

Superior 上段

Lateral basal 外侧基底段

Posterior basal 后基底段

Right lung
右肺

Superior lobe
上叶

Apical 尖段

Posterior 后段

Anterior 前段

Middle lobe
中叶

Lateral 外侧段

Inferior lobe
下叶

Superior 上段

Lateral basal 外侧基底段

Posterior basal 后基底段

Posterior view
后面观

徐存一
Cunyi George Xu

肺：人体的呼吸器官，位于胸腔。

Lung: the respiratory organ of the human body, located in the chest cavity.

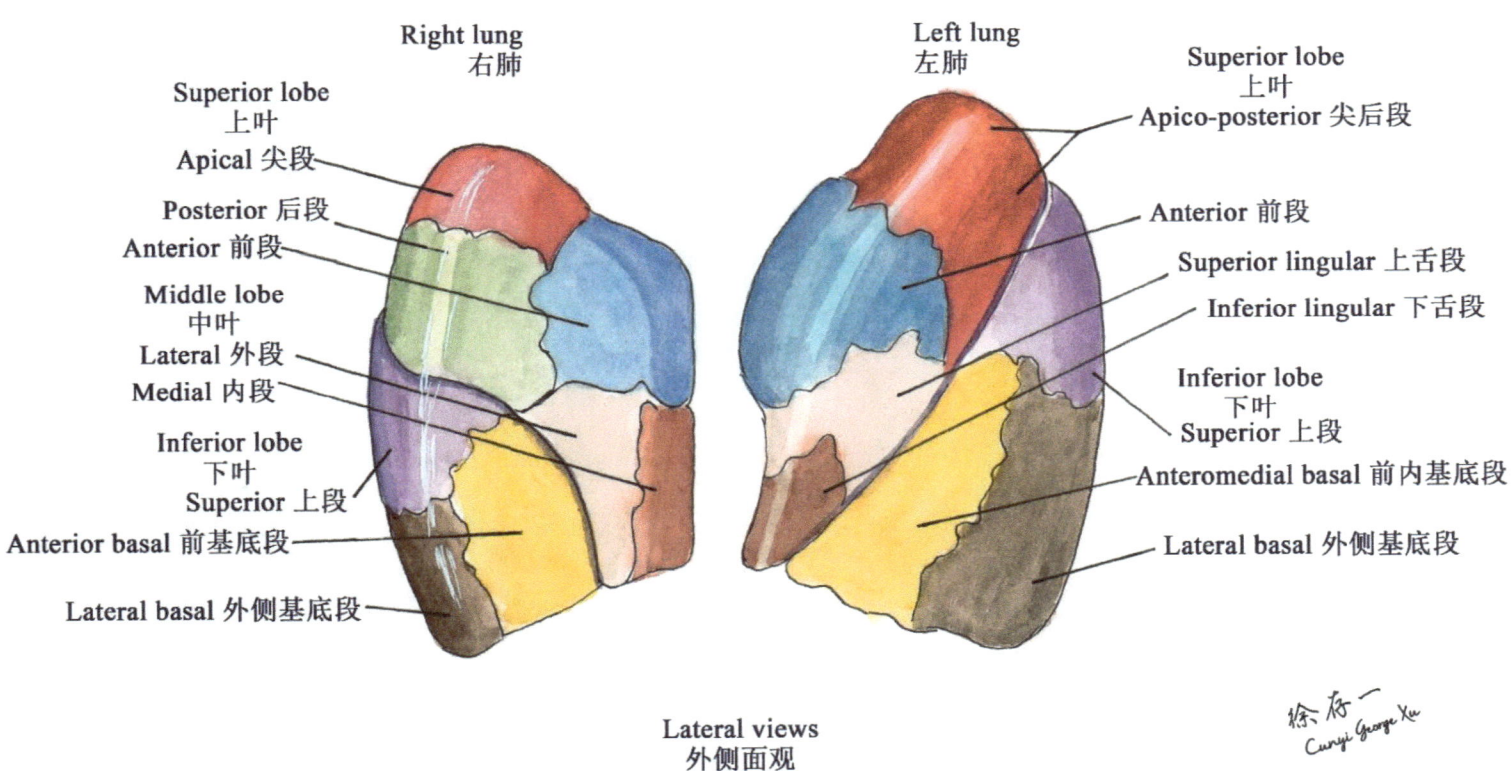

Right lung
右肺

Left lung
左肺

Superior lobe
上叶

Apical 尖段

Posterior 后段

Anterior 前段

Middle lobe
中叶

Lateral 外段

Medial 内段

Inferior lobe
下叶

Superior 上段

Anterior basal 前基底段

Lateral basal 外侧基底段

Superior lobe
上叶

Apico-posterior 尖后段

Anterior 前段

Superior lingular 上舌段

Inferior lingular 下舌段

Inferior lobe
下叶

Superior 上段

Anteromedial basal 前内基底段

Lateral basal 外侧基底段

Lateral views
外侧面观

Cunyi George Xu

肺：人体的呼吸器官，位于胸腔。

Lung: the respiratory organ of the human body, located in the chest cavity.

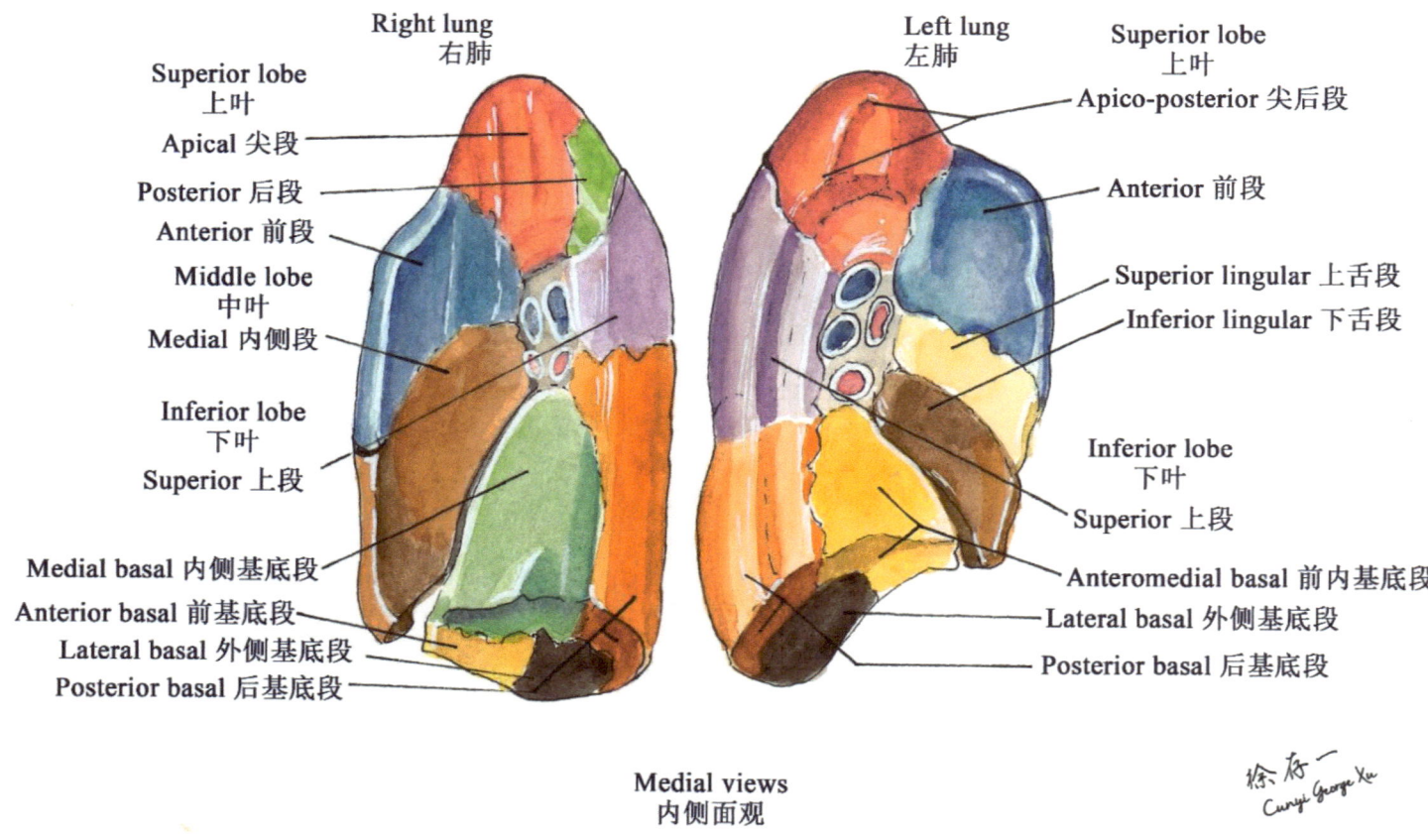

Right lung
右肺

Superior lobe
上叶

Apical 尖段

Posterior 后段

Anterior 前段

Middle lobe
中叶

Medial 内侧段

Inferior lobe
下叶

Superior 上段

Medial basal 内侧基底段

Anterior basal 前基底段

Lateral basal 外侧基底段

Posterior basal 后基底段

Left lung
左肺

Superior lobe
上叶

Apico-posterior 尖后段

Anterior 前段

Superior lingular 上舌段

Inferior lingular 下舌段

Inferior lobe
下叶

Superior 上段

Anteromedial basal 前内基底段

Lateral basal 外侧基底段

Posterior basal 后基底段

Medial views
内侧面观

Cunyi George Xu

肺：人体的呼吸器官，位于胸腔。

Lung: the respiratory organ of the human body, located in the chest cavity.

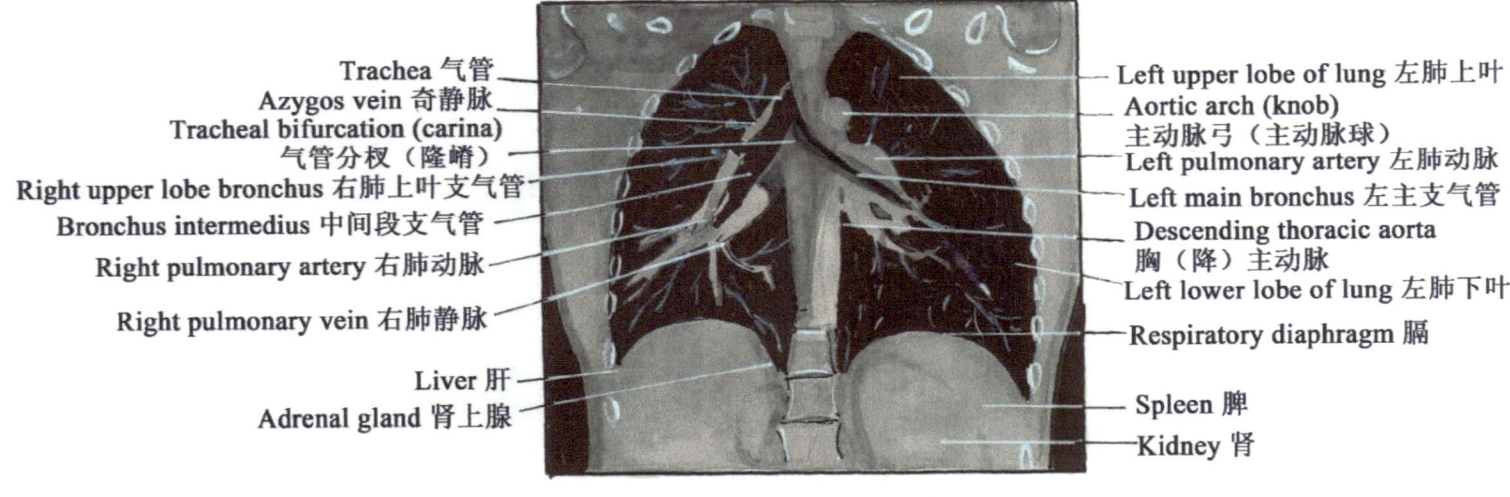

Trachea 气管
Azygos vein 奇静脉
Tracheal bifurcation (carina)
气管分杈（隆嵴）
Right upper lobe bronchus 右肺上叶支气管
Bronchus intermedius 中间段支气管
Right pulmonary artery 右肺动脉
Right pulmonary vein 右肺静脉
Liver 肝
Adrenal gland 肾上腺

Left upper lobe of lung 左肺上叶
Aortic arch (knob)
主动脉弓（主动脉球）
Left pulmonary artery 左肺动脉
Left main bronchus 左主支气管
Descending thoracic aorta
胸（降）主动脉
Left lower lobe of lung 左肺下叶
Respiratory diaphragm 膈
Spleen 脾
Kidney 肾

Contrast windowed to accentuate lungs and bones: schema
增强窗以突出肺及骨：示意图

徐存一
Cunyé George Xu

胸部：胸的上界为颈部下界，下界为骨性胸廓下口，外界为三角肌前后缘，是人体第二大体腔局部。

Chest: the upper boundary of the chest is the lower boundary of the neck, the lower boundary is the inferior opening of the rib cage, and the outer boundary is the anterior and posterior edges of the deltoid muscle. It is the second largest body cavity of the human body.

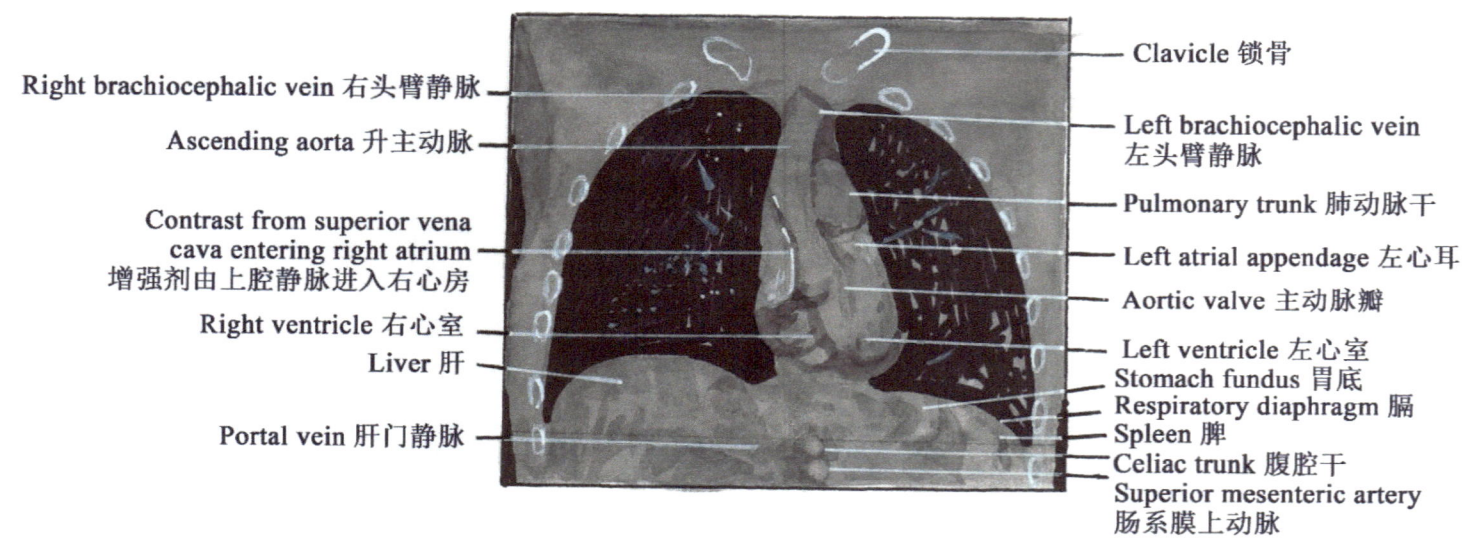

Right brachiocephalic vein 右头臂静脉

Ascending aorta 升主动脉

Contrast from superior vena cava entering right atrium 增强剂由上腔静脉进入右心房

Right ventricle 右心室

Liver 肝

Portal vein 肝门静脉

Clavicle 锁骨

Left brachiocephalic vein 左头臂静脉

Pulmonary trunk 肺动脉干

Left atrial appendage 左心耳

Aortic valve 主动脉瓣

Left ventricle 左心室

Stomach fundus 胃底

Respiratory diaphragm 膈

Spleen 脾

Celiac trunk 腹腔干

Superior mesenteric artery 肠系膜上动脉

Contrast windowed to accentuate lungs and bones: schema
增强窗以突出肺及骨：示意图

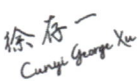

胸部：胸的上界为颈部下界，下界为骨性胸廓下口，外界为三角肌前后缘，是人体第二大体腔局部。

Chest: the upper boundary of the chest is the lower boundary of the neck, the lower boundary is the inferior opening of the rib cage, and the outer boundary is the anterior and posterior edges of the deltoid muscle. It is the second largest body cavity of the human body.

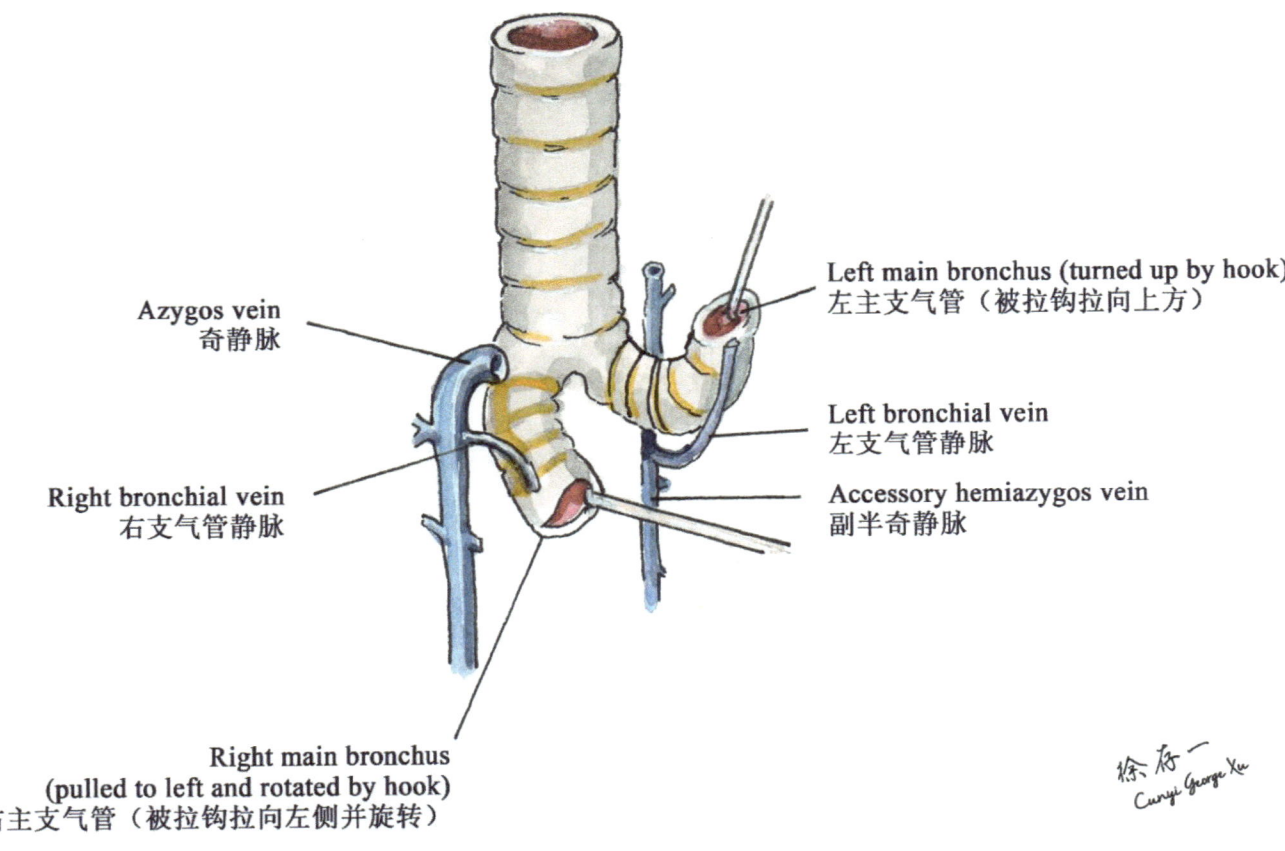

Azygos vein
奇静脉

Right bronchial vein
右支气管静脉

Right main bronchus
(pulled to left and rotated by hook)
右主支气管（被拉钩拉向左侧并旋转）

Left main bronchus (turned up by hook)
左主支气管（被拉钩拉向上方）

Left bronchial vein
左支气管静脉

Accessory hemiazygos vein
副半奇静脉

支气管静脉：收集从肺部及气道回流的血液，带走废物和二氧化碳等代谢废物。

Bronchial veins: collect blood returning from the lungs and airways, carrying away waste and metabolic by-products such as carbon dioxide.

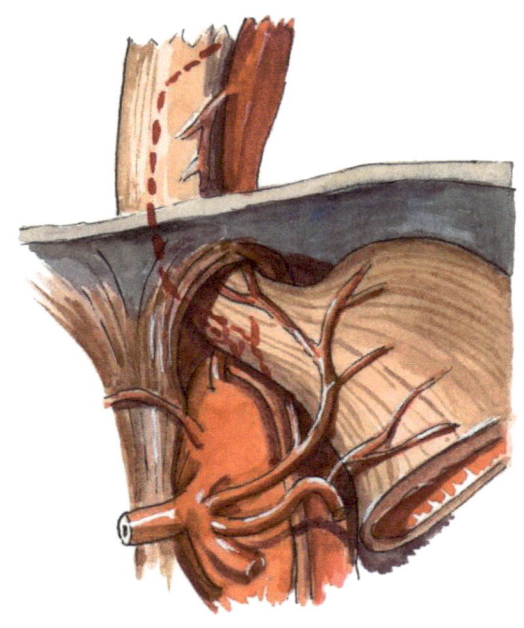

Common variations of esophageal branches: esophageal branches may originate from left inferior phrenic artery and/or directly from celiac trunk. Branches to abdominal esophagus may also come from splenic or short gastric arteries.

食管动脉常见变异：食管的动脉支可能起源于左膈下动脉和/或直接来自腹腔干。腹段的食管动脉支也可能来自脾动脉或胃短动脉。

徐存一
Cunyi George Xu

食管动脉：是动脉血流、营养物质运输和营养物质的重要来源之一，是食管内壁支撑和食管营养物质的重要供应者。

Esophageal artery: one of the important sources of arterial blood flow, transporting nutrients, and playing a crucial role in supplying the inner wall and nutrients to the esophagus.

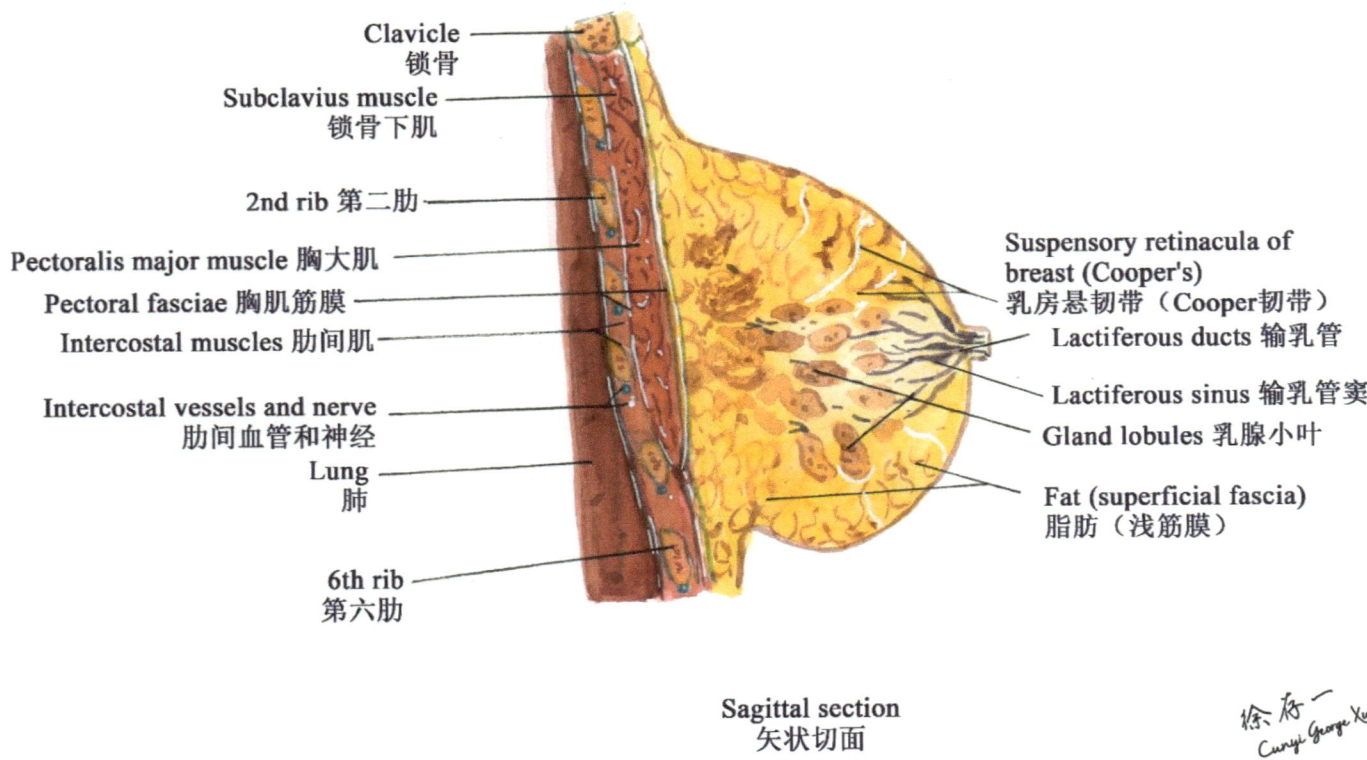

Clavicle 锁骨

Subclavius muscle 锁骨下肌

2nd rib 第二肋

Pectoralis major muscle 胸大肌

Pectoral fasciae 胸肌筋膜

Intercostal muscles 肋间肌

Intercostal vessels and nerve 肋间血管和神经

Lung 肺

6th rib 第六肋

Suspensory retinacula of breast (Cooper's) 乳房悬韧带（Cooper韧带）

Lactiferous ducts 输乳管

Lactiferous sinus 输乳管窦

Gland lobules 乳腺小叶

Fat (superficial fascia) 脂肪（浅筋膜）

Sagittal section 矢状切面

乳腺：分泌乳汁、保护乳房、维持内分泌平衡等作用。

Mammary gland: secretes milk, protects the breast, maintains endocrine balance, and performs other functions.

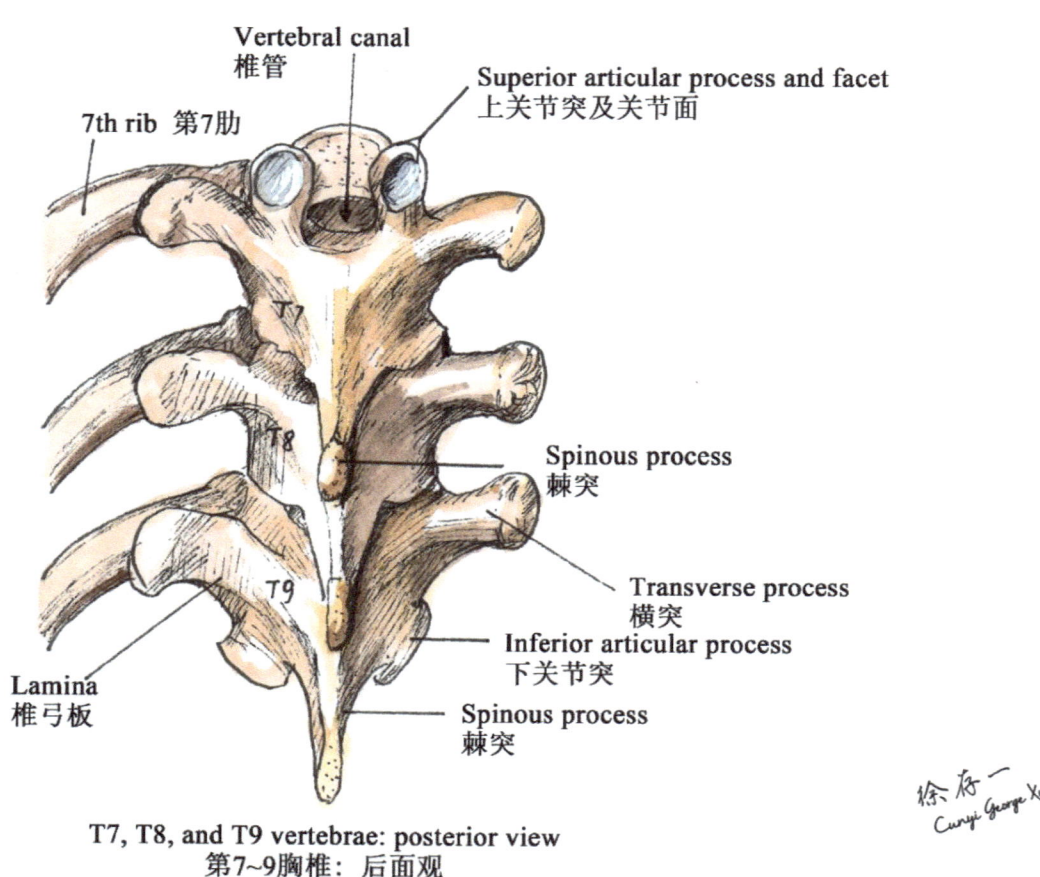

Vertebral canal
椎管

Superior articular process and facet
上关节突及关节面

7th rib 第7肋

T7

T8

Spinous process
棘突

Transverse process
横突

Inferior articular process
下关节突

T9

Lamina
椎弓板

Spinous process
棘突

徐存一
Cunyi George Xu

T7, T8, and T9 vertebrae: posterior view
第7~9胸椎：后面观

第 7 至第 9 胸椎：支撑脊柱、保护脊髓，提供肌肉和韧带附着点。

T7, T8, and T9 vertebrae: support the spine, protect the spinal cord, and provide attachment points for muscles and ligaments.

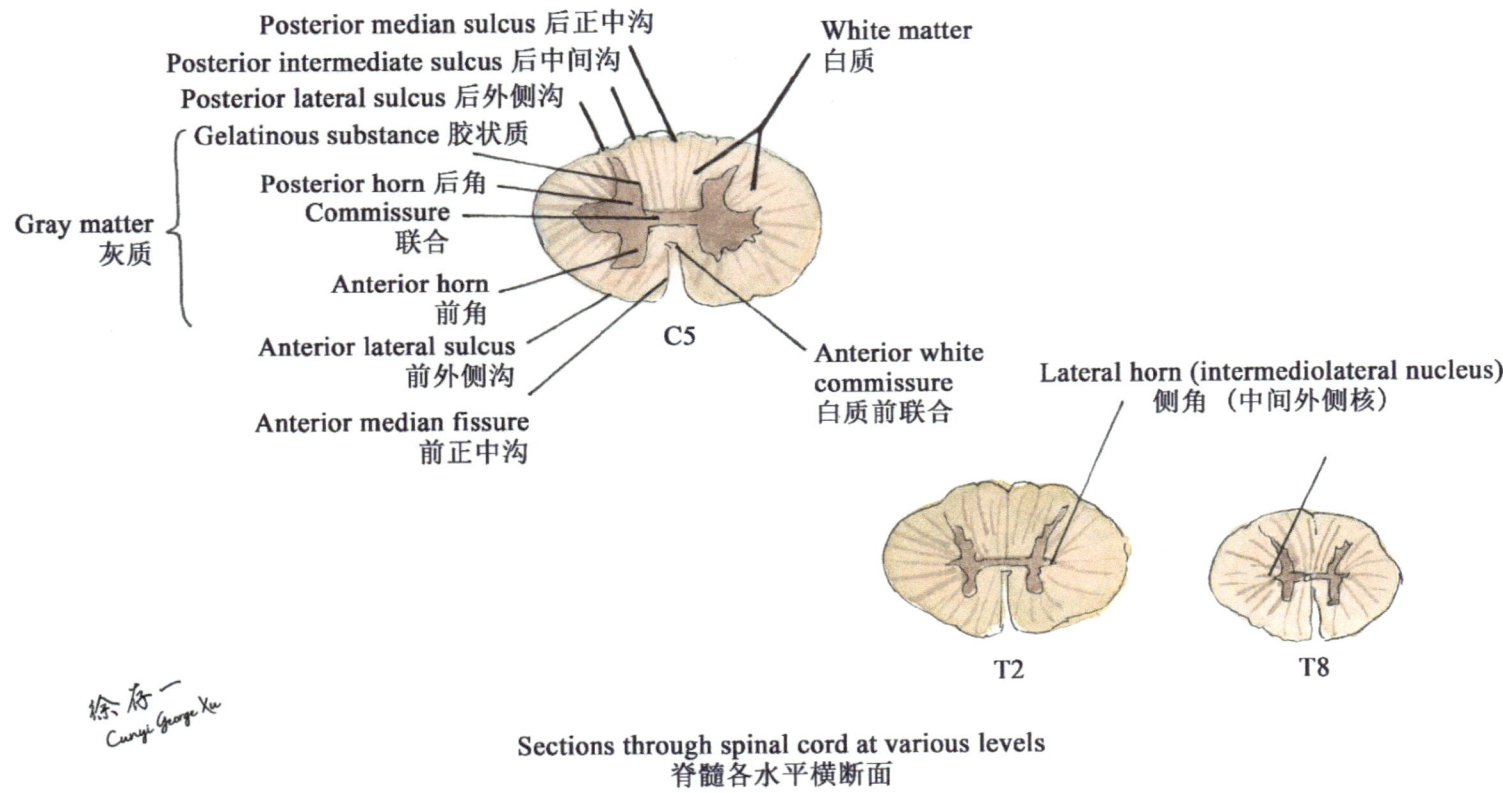

Posterior median sulcus 后正中沟
Posterior intermediate sulcus 后中间沟
Posterior lateral sulcus 后外侧沟
Gelatinous substance 胶状质
Posterior horn 后角
Commissure 联合
Anterior horn 前角
Anterior lateral sulcus 前外侧沟
Anterior median fissure 前正中沟
Gray matter 灰质
White matter 白质
Anterior white commissure 白质前联合
C5

Lateral horn (intermediolateral nucleus) 侧角 （中间外侧核）
T2
T8

Cunyi George Xu

Sections through spinal cord at various levels
脊髓各水平横断面

脊髓： 传导、反射、控制身体活动、维持机体热能、支配肌肉收缩和腺体分泌等。

Spinal cord: conduction, reflex actions, control of body activity, maintenance of body heat, and regulation of muscle contraction and glandular secretion.

初级人体解剖图谱

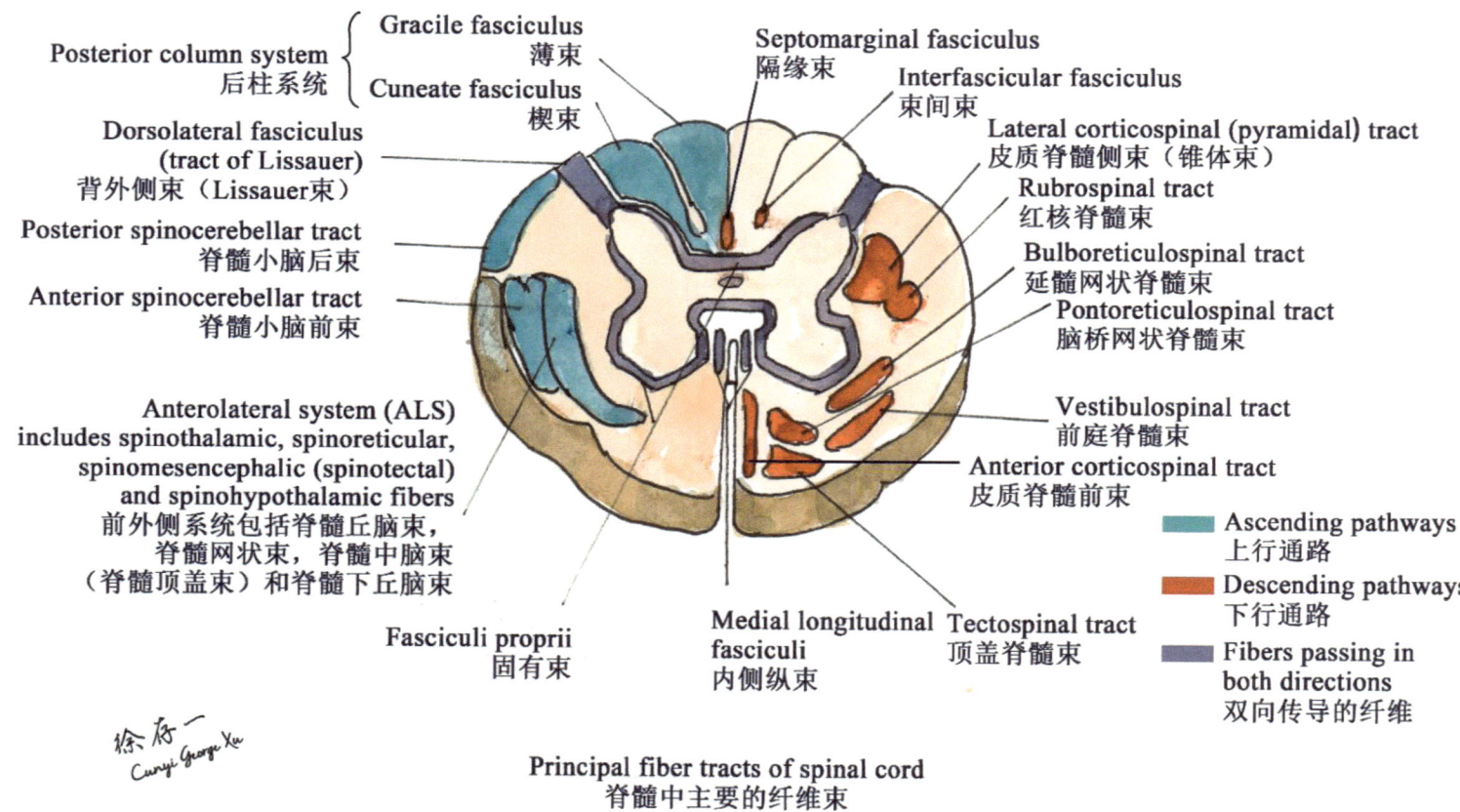

Posterior column system
后柱系统
Gracile fasciculus
薄束
Cuneate fasciculus
楔束

Septomarginal fasciculus
隔缘束
Interfascicular fasciculus
束间束
Lateral corticospinal (pyramidal) tract
皮质脊髓侧束（锥体束）
Rubrospinal tract
红核脊髓束
Bulboreticulospinal tract
延髓网状脊髓束
Pontoreticulospinal tract
脑桥网状脊髓束

Dorsolateral fasciculus
(tract of Lissauer)
背外侧束（Lissauer束）
Posterior spinocerebellar tract
脊髓小脑后束
Anterior spinocerebellar tract
脊髓小脑前束

Vestibulospinal tract
前庭脊髓束
Anterior corticospinal tract
皮质脊髓前束

Anterolateral system (ALS)
includes spinothalamic, spinoreticular,
spinomesencephalic (spinotectal)
and spinohypothalamic fibers
前外侧系统包括脊髓丘脑束，
脊髓网状束，脊髓中脑束
（脊髓顶盖束）和脊髓下丘脑束

Ascending pathways
上行通路
Descending pathways
下行通路
Fibers passing in
both directions
双向传导的纤维

Fasciculi proprii
固有束
Medial longitudinal
fasciculi
内侧纵束
Tectospinal tract
顶盖脊髓束

Principal fiber tracts of spinal cord
脊髓中主要的纤维束

脊髓：传导、反射、控制身体活动、维持机体热能、支配肌肉收缩和腺体分泌等。

Spinal cord: conduction, reflex actions, control of body activity, maintenance of body heat, and regulation of muscle contraction and glandular secretion.

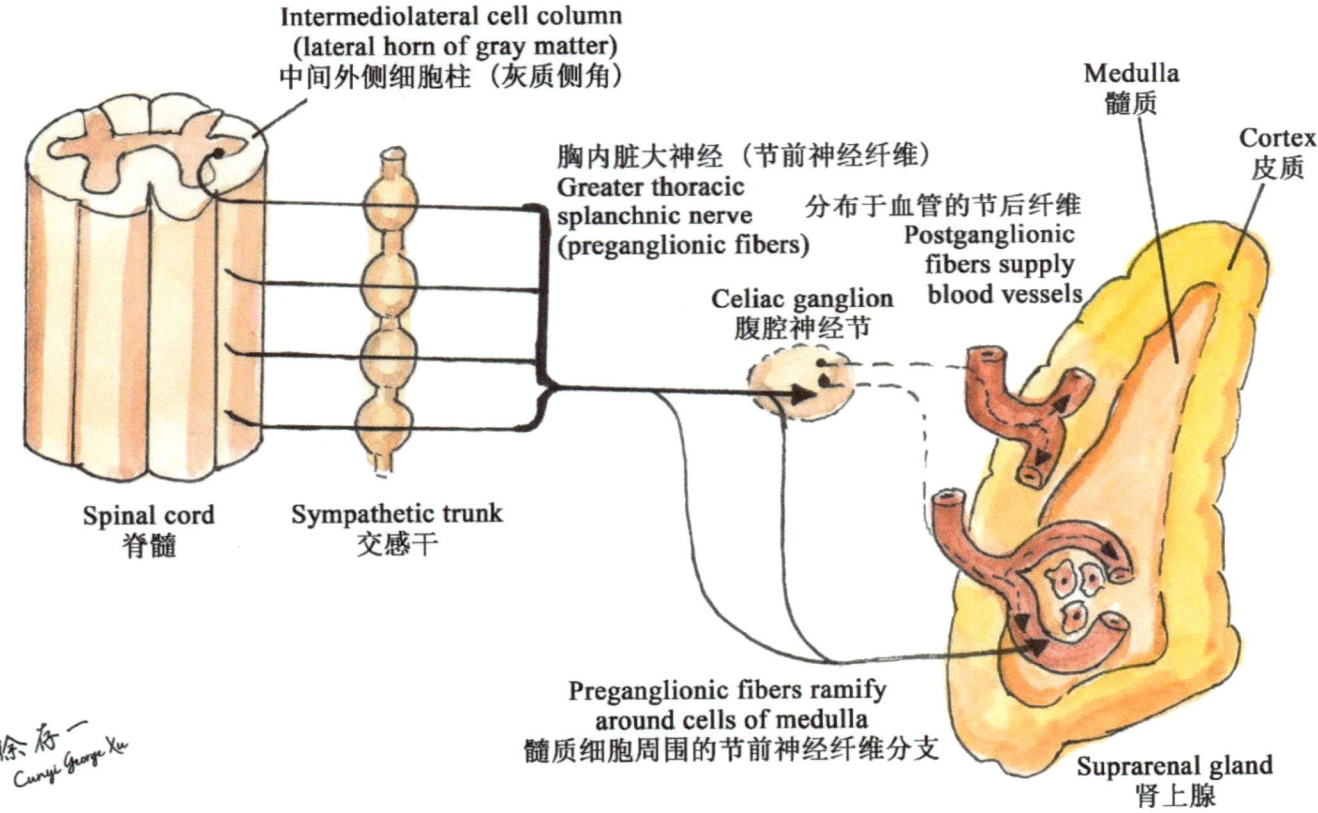

Intermediolateral cell column
(lateral horn of gray matter)
中间外侧细胞柱（灰质侧角）

胸内脏大神经（节前神经纤维）
Greater thoracic
splanchnic nerve
(preganglionic fibers)

分布于血管的节后纤维
Postganglionic
fibers supply
blood vessels

Medulla
髓质

Cortex
皮质

Celiac ganglion
腹腔神经节

Spinal cord
脊髓

Sympathetic trunk
交感干

Preganglionic fibers ramify
around cells of medulla
髓质细胞周围的节前神经纤维分支

Suprarenal gland
肾上腺

徐存一
Cunyi George Xu

肾上腺自主神经：交感神经支配肾上腺，刺激肾上腺分泌肾上腺素等应激激素，帮助应对压力。

The autonomic nervous system of the suprarenal gland: the sympathetic nervous system innervates the suprarenal gland, stimulating it to secrete stress hormones, such as adrenaline, to help cope with stress.

初级人体解剖图谱

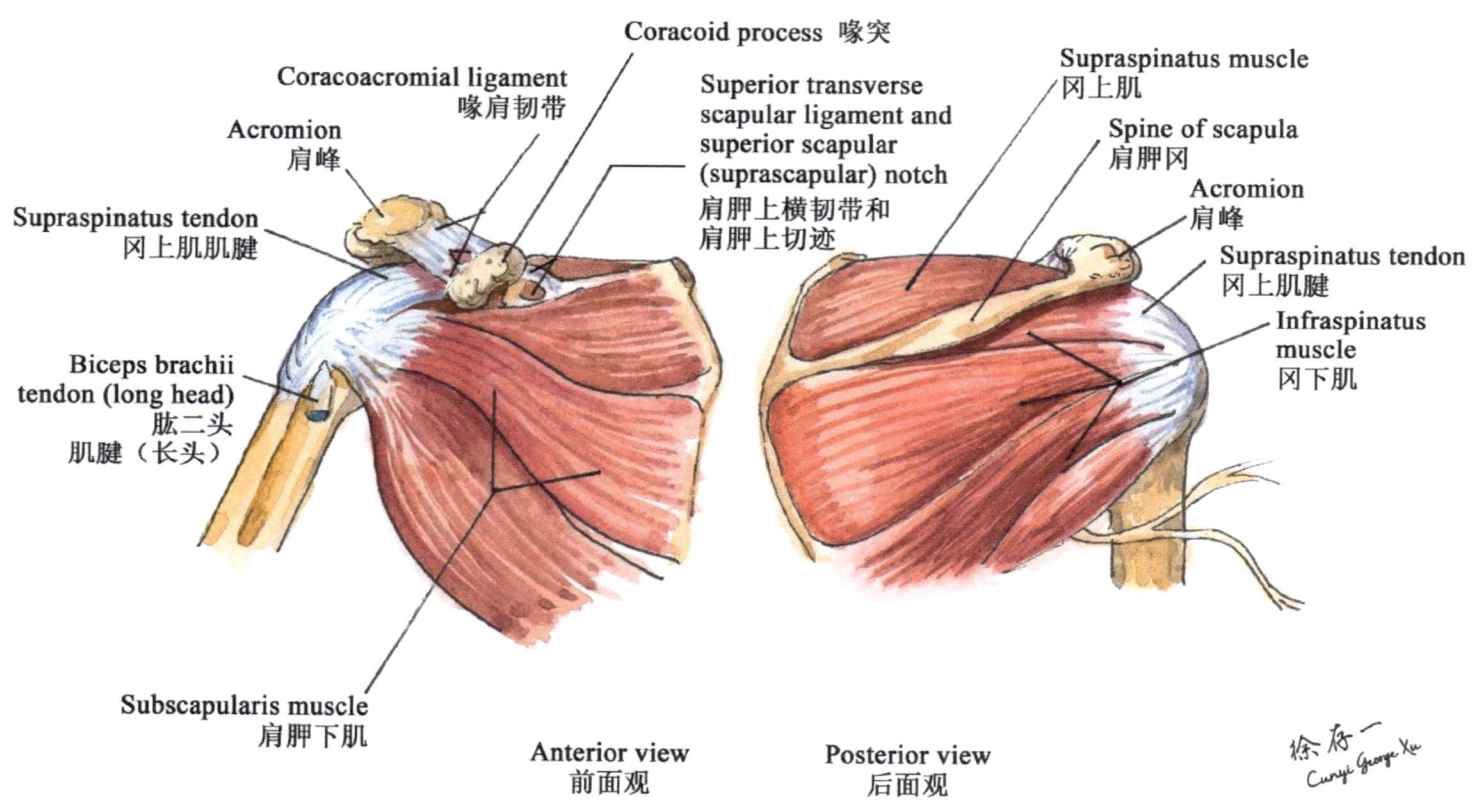

Coracoid process 喙突

Coracoacromial ligament
喙肩韧带

Acromion
肩峰

Superior transverse
scapular ligament and
superior scapular
(suprascapular) notch
肩胛上横韧带和
肩胛上切迹

Supraspinatus muscle
冈上肌

Spine of scapula
肩胛冈

Acromion
肩峰

Supraspinatus tendon
冈上肌肌腱

Supraspinatus tendon
冈上肌腱

Infraspinatus
muscle
冈下肌

Biceps brachii
tendon (long head)
肱二头
肌腱（长头）

Subscapularis muscle
肩胛下肌

Anterior view
前面观

Posterior view
后面观

徐存一
Cunyi George Xu

肩袖肌群：肩部的运动、稳定性和力量传递。

Rotator cuff muscle group: responsible for shoulder movement, stability, and power transmission.

第四章　腹部和盆部

Chapter 4 Abdomen and Pelvis

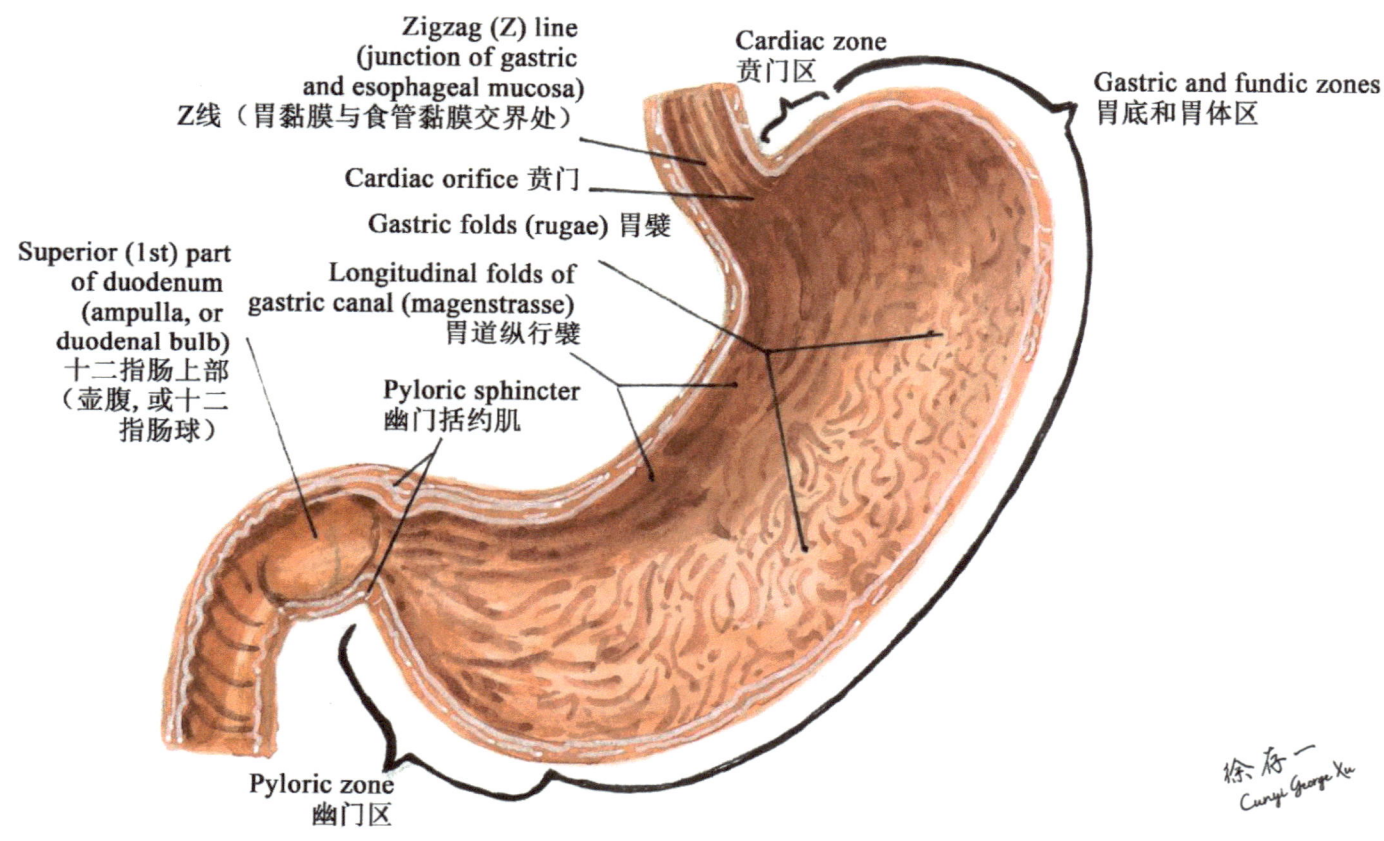

Zigzag (Z) line
(junction of gastric
and esophageal mucosa)
Z线（胃黏膜与食管黏膜交界处）

Cardiac zone
贲门区

Gastric and fundic zones
胃底和胃体区

Cardiac orifice 贲门

Gastric folds (rugae) 胃襞

Superior (1st) part
of duodenum
(ampulla, or
duodenal bulb)
十二指肠上部
（壶腹, 或十二
指肠球）

Longitudinal folds of
gastric canal (magenstrasse)
胃道纵行襞

Pyloric sphincter
幽门括约肌

Pyloric zone
幽门区

徐存一
Cunyi George Xu

胃：储存和消化食物。

Stomach: stores and digests food.

Outer longitudinal muscle layer (with window cut)
外层纵行肌层（开窗）

Inner circular muscle layer (with window cut)
内层环形肌层（开窗）

Submucosa with duodenal (Brunner's) glands
黏膜下层中的十二指肠腺（Brunner腺）

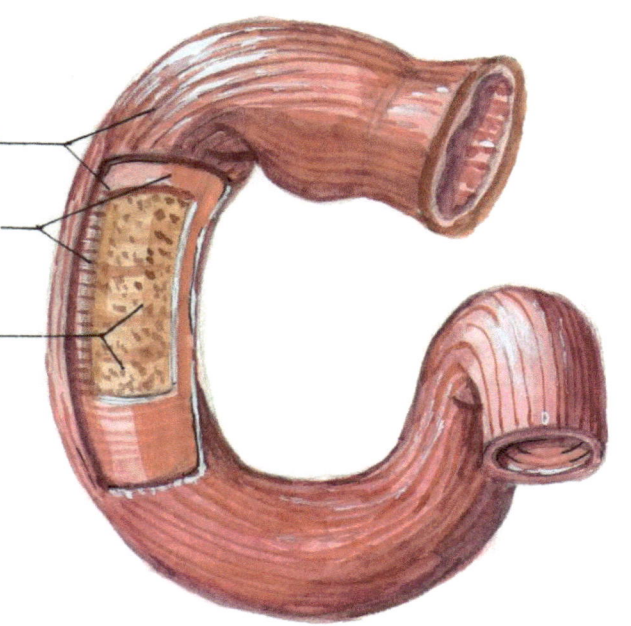

十二指肠：促进食物的消化和吸收。

Duodenum: promotes the digestion and absorption of food.

初级人体解剖图谱

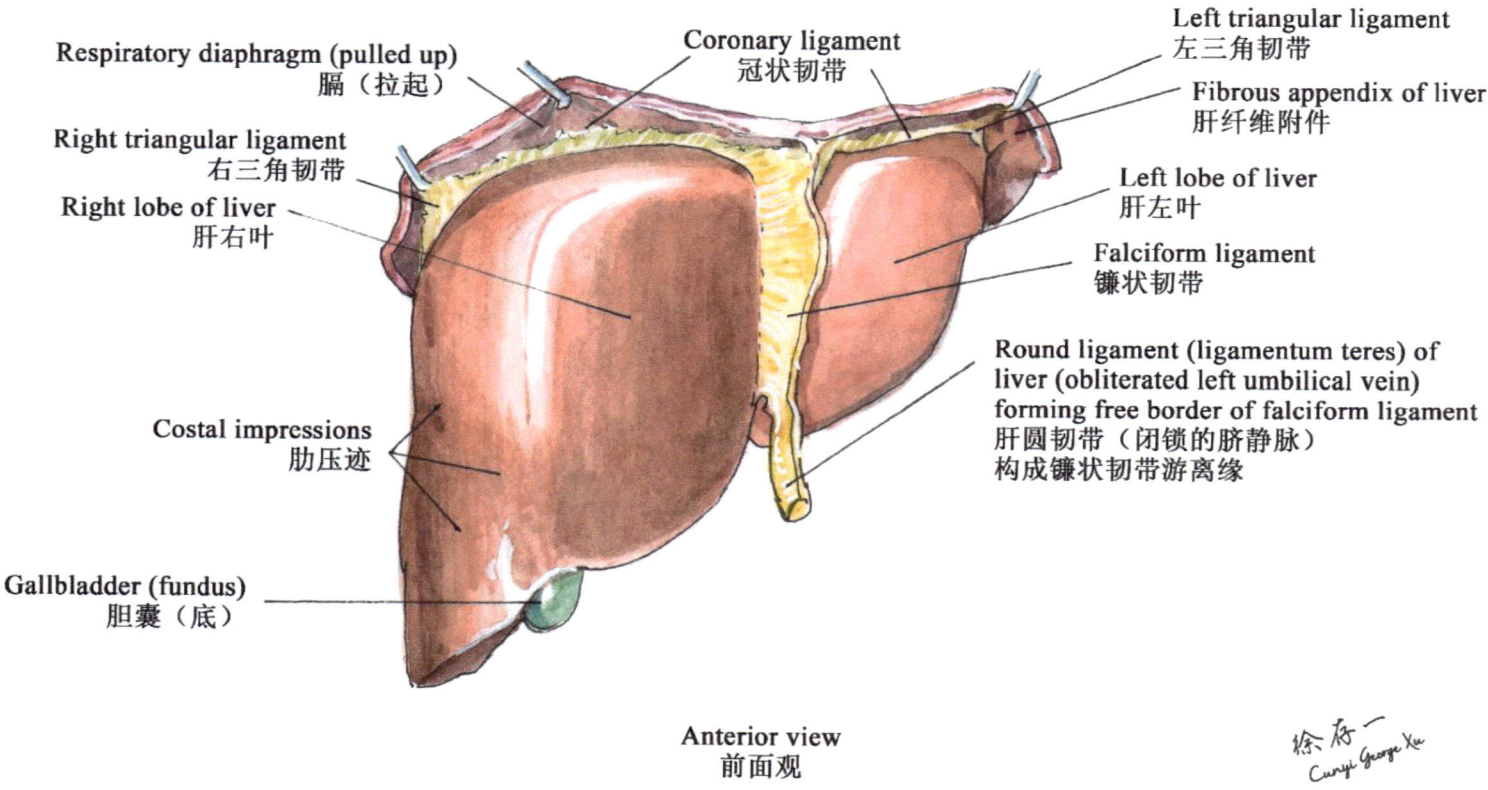

Respiratory diaphragm (pulled up)
膈（拉起）

Coronary ligament
冠状韧带

Left triangular ligament
左三角韧带

Fibrous appendix of liver
肝纤维附件

Right triangular ligament
右三角韧带

Right lobe of liver
肝右叶

Left lobe of liver
肝左叶

Falciform ligament
镰状韧带

Round ligament (ligamentum teres) of liver (obliterated left umbilical vein) forming free border of falciform ligament
肝圆韧带（闭锁的脐静脉）构成镰状韧带游离缘

Costal impressions
肋压迹

Gallbladder (fundus)
胆囊（底）

Anterior view
前面观

Cunyi George Xu

肝脏：解毒、合成蛋白质、储存能量、分泌胆汁以及参与代谢。

Liver: detoxification, protein synthesis, energy storage, bile secretion, and participation in metabolism.

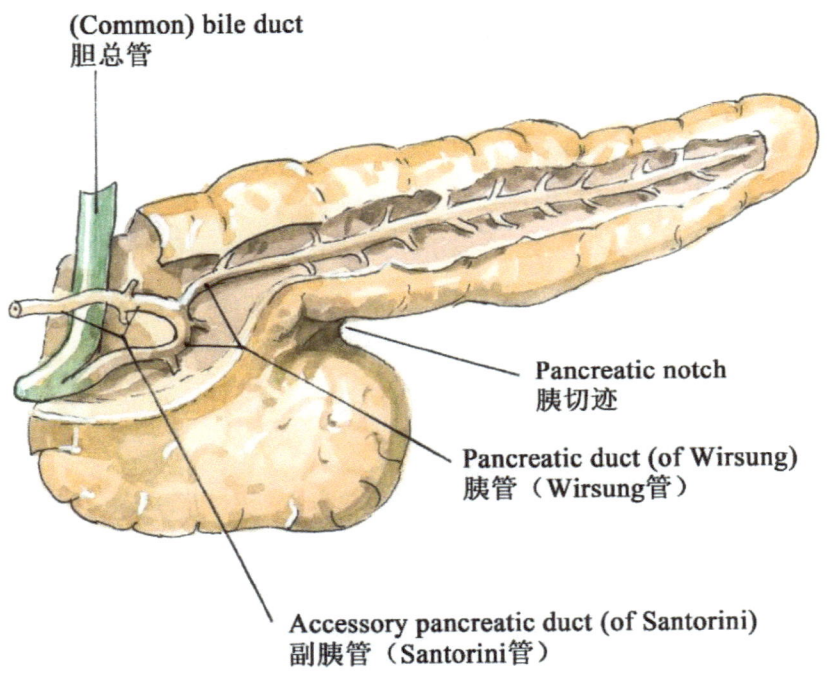

(Common) bile duct
胆总管

Pancreatic notch
胰切迹

Pancreatic duct (of Wirsung)
胰管（Wirsung管）

Accessory pancreatic duct (of Santorini)
副胰管（Santorini管）

胰腺：通过胰管输送消化酶到小肠以分解食物，帮助消化。

Pancreas: delivers digestive enzymes to the small intestine through the pancreatic duct to break down food and aid in digestion.

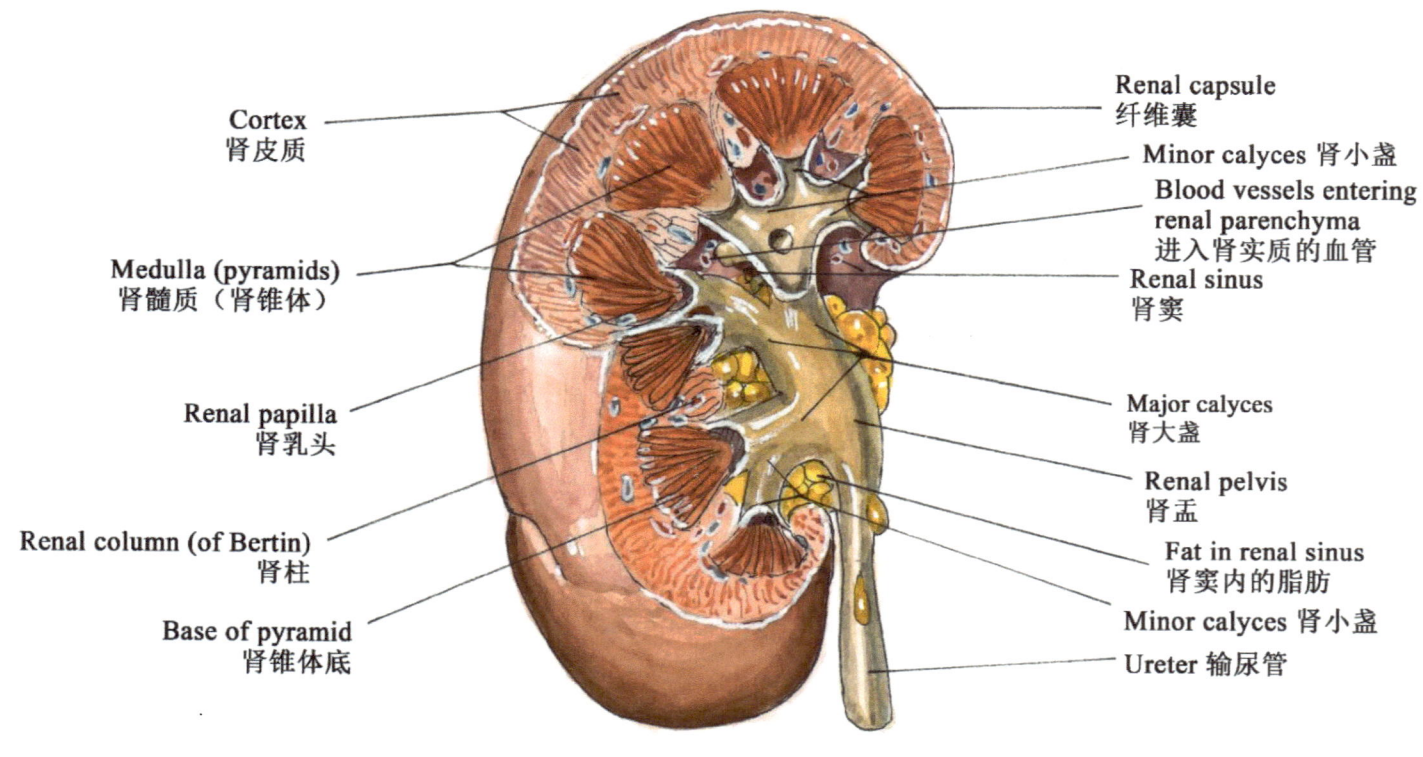

Cortex
肾皮质

Medulla (pyramids)
肾髓质（肾锥体）

Renal papilla
肾乳头

Renal column (of Bertin)
肾柱

Base of pyramid
肾锥体底

Renal capsule
纤维囊

Minor calyces 肾小盏

Blood vessels entering
renal parenchyma
进入肾实质的血管

Renal sinus
肾窦

Major calyces
肾大盏

Renal pelvis
肾盂

Fat in renal sinus
肾窦内的脂肪

Minor calyces 肾小盏

Ureter 输尿管

Right kidney sectioned in several planes, exposing parenchyma and renal pelvis
右肾的不同断面，暴露肾实质和肾盂

徐存一
Cunye George Xu

肾脏：过滤血液，排除废物和多余水分，保持身体内环境的稳定。

Kidney: filters blood, eliminates waste and excess water, and maintains a stable internal environment in the body.

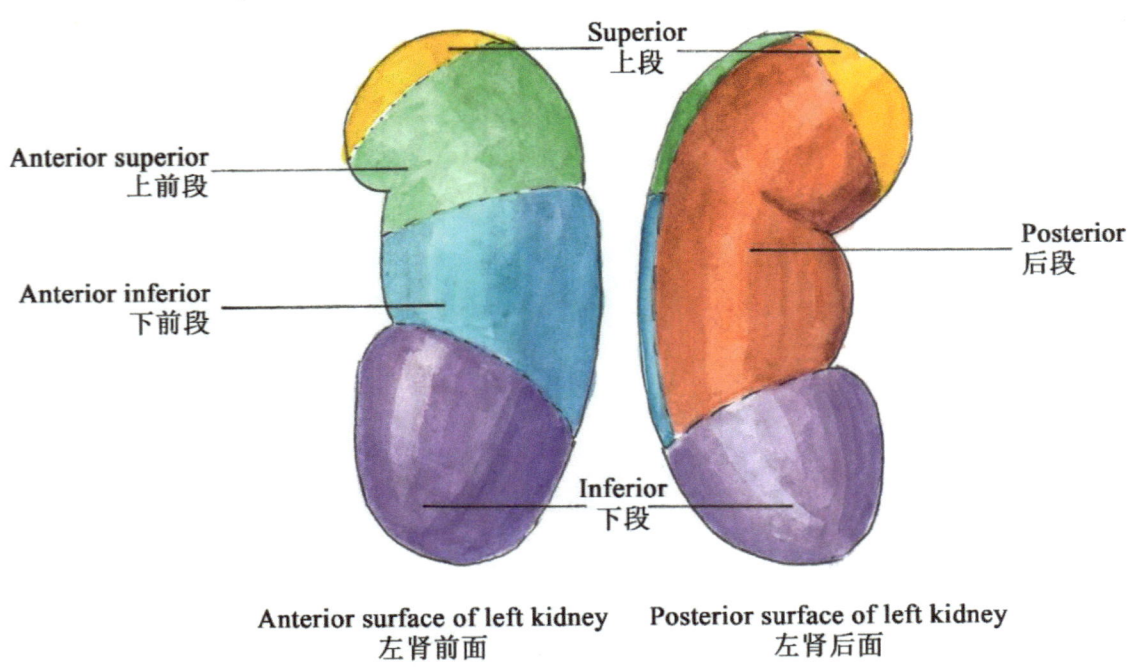

Superior
上段

Anterior superior
上前段

Anterior inferior
下前段

Posterior
后段

Inferior
下段

Anterior surface of left kidney
左肾前面

Posterior surface of left kidney
左肾后面

Vascular renal segments
肾血管分段

肾血管: 肾血管分段方式在肾脏手术中至关重要，有助于精确切除病变区域，保留健康部分。

Vessels of the kidney: the segmentation of renal vasculature is crucial in kidney surgery, helping to precisely remove diseased areas while preserving healthy parts.

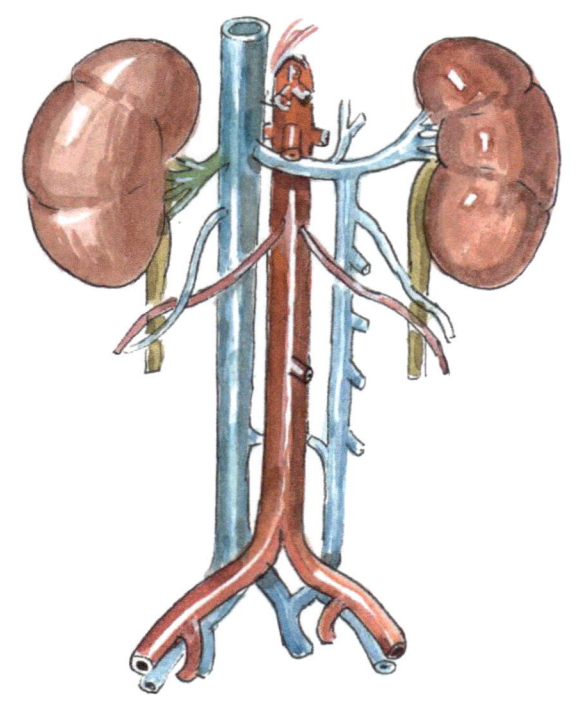

Persistent left inferior vena cava may join left renal vein
延续性左下腔静脉可与左肾静脉汇合

肾动静脉：肾动脉为肾脏提供血液供应，以满足其代谢需求；肾静脉则负责将含有废物和多余液体的血液从肾脏运输出去。

Renal arteries and veins: renal arteries provide the blood supply to the kidneys to meet their metabolic needs, while renal veins carry blood containing waste and excess fluid away from the kidneys.

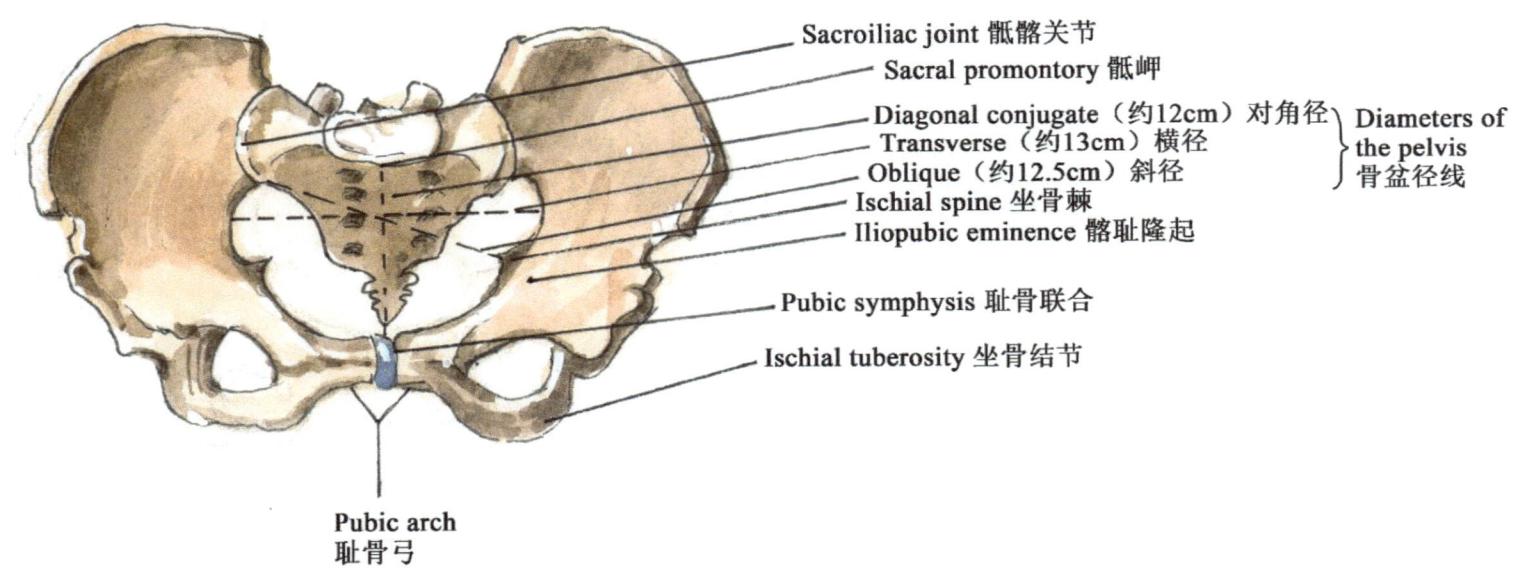

Sacroiliac joint 骶髂关节

Sacral promontory 骶岬

Diagonal conjugate （约12cm）对角径

Transverse （约13cm）横径

Oblique （约12.5cm）斜径

Ischial spine 坐骨棘

Iliopubic eminence 髂耻隆起

Pubic symphysis 耻骨联合

Ischial tuberosity 坐骨结节

Diameters of the pelvis 骨盆径线

Pubic arch
耻骨弓

Female pelvis/female pelvic inlet: anterior view
女性骨盆/女性骨盆入口：前面观

女性骨盆：稳定骨盆、承重以及帮助胎儿在分娩过程中顺利通过产道。

Female pelvis: stabilises the pelvis, supports weight, and assists the fetus in smoothly passing through the birth canal during delivery.

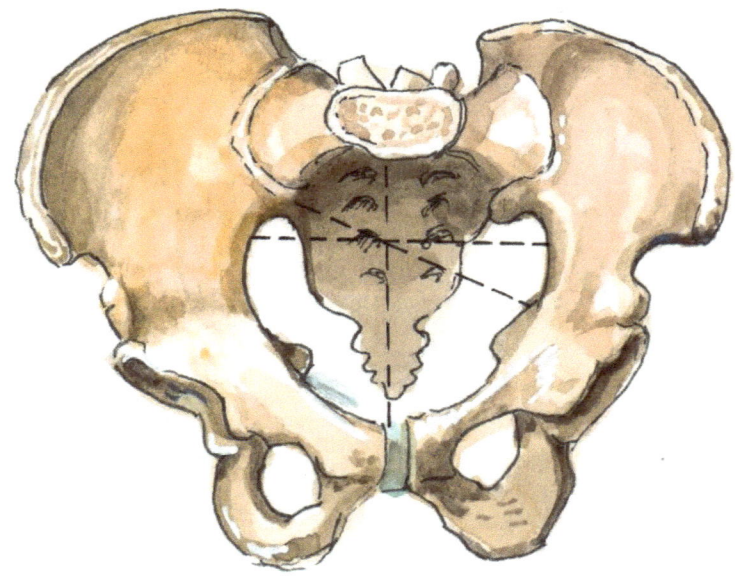

All measurements slightly shorter in relation
to body size than in female.
Pelvic inlet oriented more antero-posteriorly
than in female, where it tends to be transversely
oval. Pubic symphysis deeper (taller). Pubic arch
(subpubic angle) narrower. Ischial tuberosities
less far apart. Iliac wings less flared.

男性所有径线均较相似体形女性短,
骨盆入口较女性更为前倾,是横卵圆形。
耻骨联合更深(更高),耻骨弓(耻骨下角)更窄。
坐骨结节相距更近,髂骨翼更加聚拢。

Male pelvis/male pelvic inlet: anterior view
男性骨盆/男性骨盆入口:前面观

男性骨盆: 提供身体支撑和稳定性,适应较大的体重承载和力量传递。

Male pelvis: provides body support and stability, adapted for carrying greater body weight and transferring force.

第五章　四肢

Chapter 5　Limbs

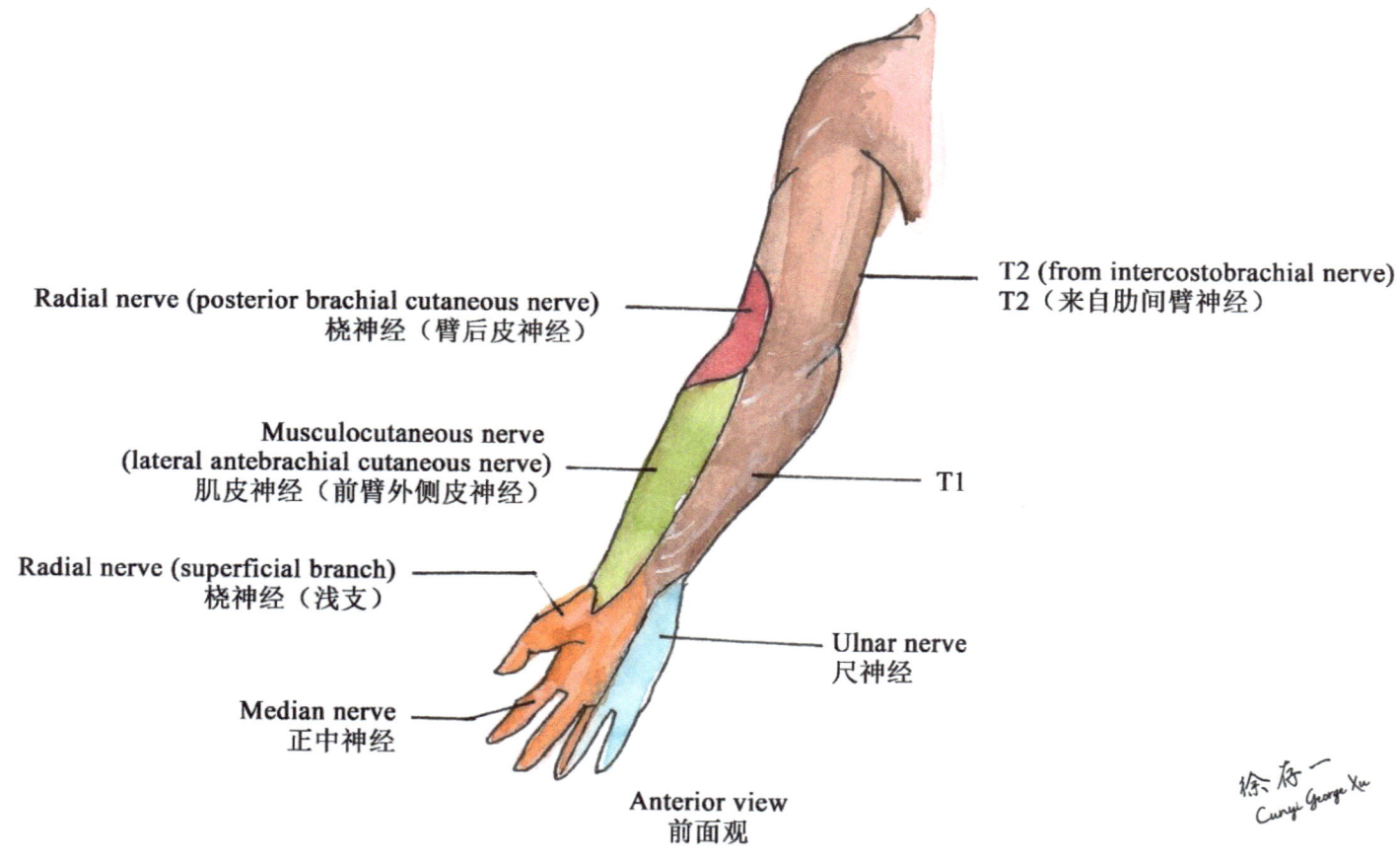

Radial nerve (posterior brachial cutaneous nerve)
桡神经（臂后皮神经）

T2 (from intercostobrachial nerve)
T2（来自肋间臂神经）

Musculocutaneous nerve
(lateral antebrachial cutaneous nerve)
肌皮神经（前臂外侧皮神经）

T1

Radial nerve (superficial branch)
桡神经（浅支）

Ulnar nerve
尺神经

Median nerve
正中神经

Anterior view
前面观

上肢的神经：控制手臂、手腕和手指的运动，传递感觉信息到大脑。

The nerves of the upper limbs: control the movement of the arms, wrists, and fingers, and transmit sensory information to the brain.

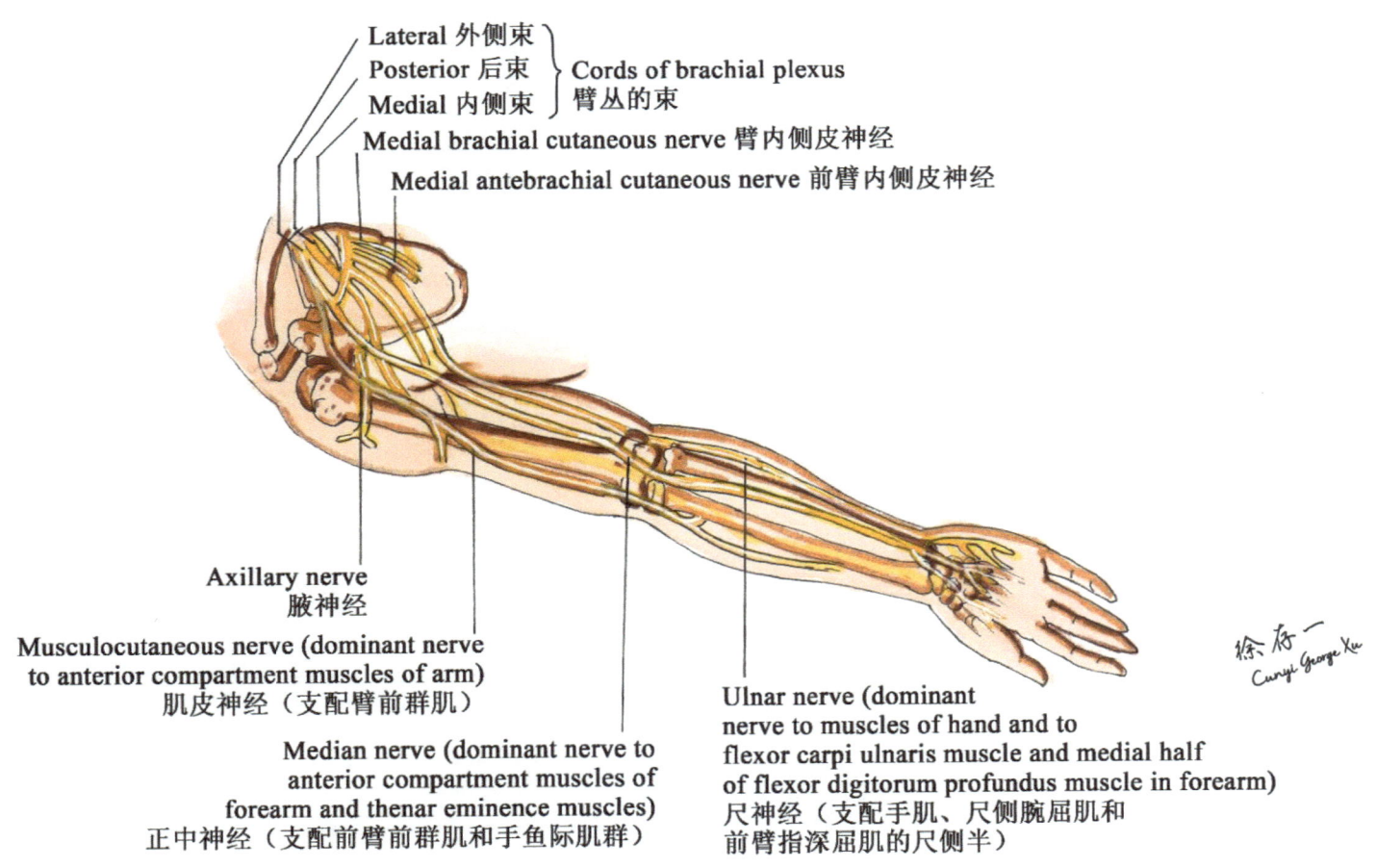

Lateral 外侧束
Posterior 后束 } Cords of brachial plexus 臂丛的束
Medial 内侧束

Medial brachial cutaneous nerve 臂内侧皮神经

Medial antebrachial cutaneous nerve 前臂内侧皮神经

Axillary nerve
腋神经

Musculocutaneous nerve (dominant nerve
to anterior compartment muscles of arm)
肌皮神经（支配臂前群肌）

Median nerve (dominant nerve to
anterior compartment muscles of
forearm and thenar eminence muscles)
正中神经（支配前臂前群肌和手鱼际肌群）

Ulnar nerve (dominant
nerve to muscles of hand and to
flexor carpi ulnaris muscle and medial half
of flexor digitorum profundus muscle in forearm)
尺神经（支配手肌、尺侧腕屈肌和
前臂指深屈肌的尺侧半）

上肢的神经：控制手臂、手腕和手指的运动，传递感觉信息到大脑。

The nerves of the upper limbs: control the movement of the arms, wrists, and fingers, and transmit sensory information to the brain.

初级人体解剖图谱

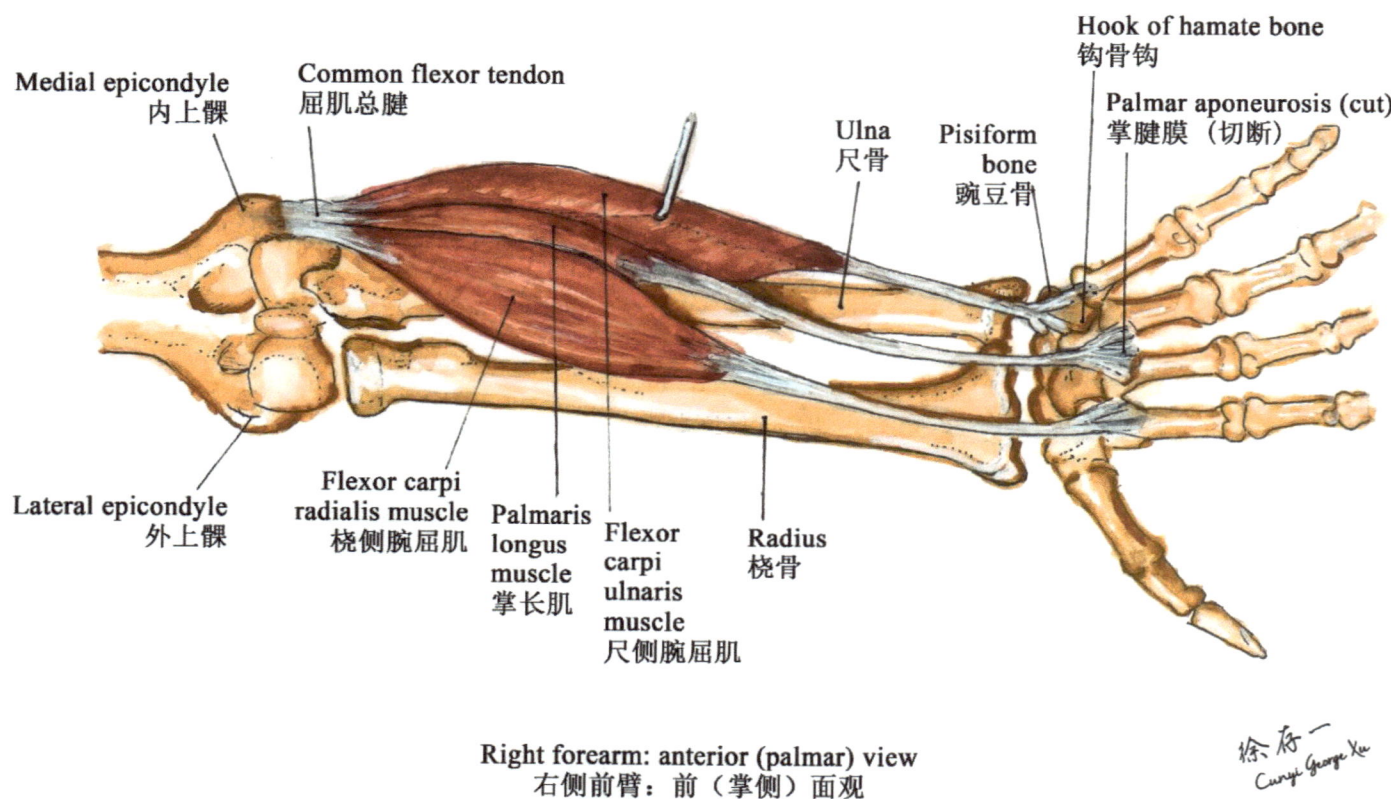

Medial epicondyle
内上髁

Common flexor tendon
屈肌总腱

Hook of hamate bone
钩骨钩

Palmar aponeurosis (cut)
掌腱膜 （切断）

Ulna
尺骨

Pisiform
bone
豌豆骨

Lateral epicondyle
外上髁

Flexor carpi
radialis muscle
桡侧腕屈肌

Palmaris
longus
muscle
掌长肌

Flexor
carpi
ulnaris
muscle
尺侧腕屈肌

Radius
桡骨

Right forearm: anterior (palmar) view
右侧前臂：前（掌侧）面观

徐存一
Cunyi George Xu

右侧前臂肌：负责手腕和手指的屈曲运动。

Muscles of the right forearm: responsible for the flexion movements of the wrist and fingers.

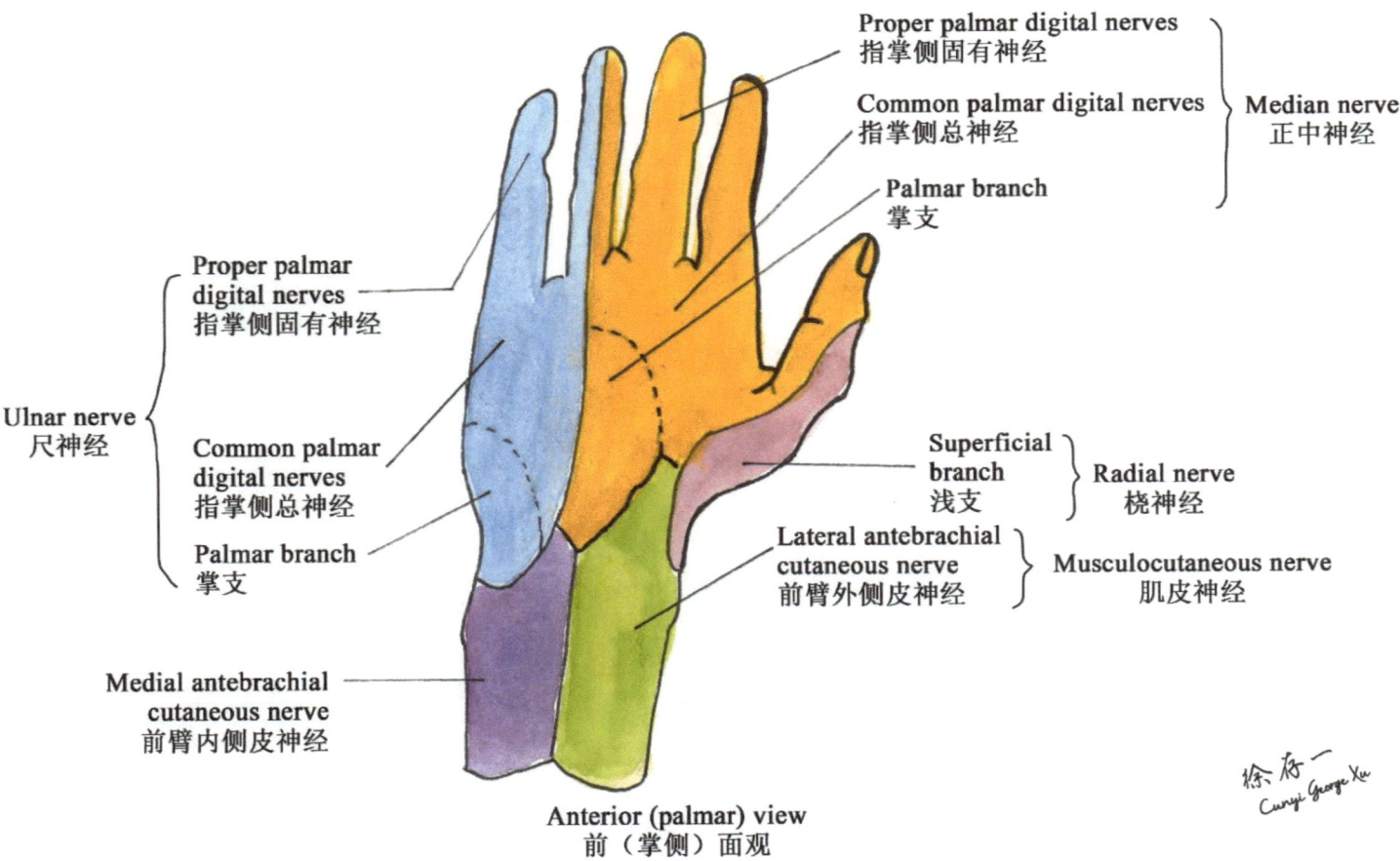

Proper palmar digital nerves
指掌侧固有神经

Common palmar digital nerves
指掌侧总神经

Palmar branch
掌支

Median nerve
正中神经

Proper palmar
digital nerves
指掌侧固有神经

Ulnar nerve
尺神经

Common palmar
digital nerves
指掌侧总神经

Palmar branch
掌支

Superficial
branch
浅支

Radial nerve
桡神经

Lateral antebrachial
cutaneous nerve
前臂外侧皮神经

Musculocutaneous nerve
肌皮神经

Medial antebrachial
cutaneous nerve
前臂内侧皮神经

Anterior (palmar) view
前（掌侧）面观

Cunyi George Xu

掌（前）神经：手掌和手指的感觉及运动控制。

Palmar (anterior) nerves: responsible for the sensation and movement control of the palms and fingers.

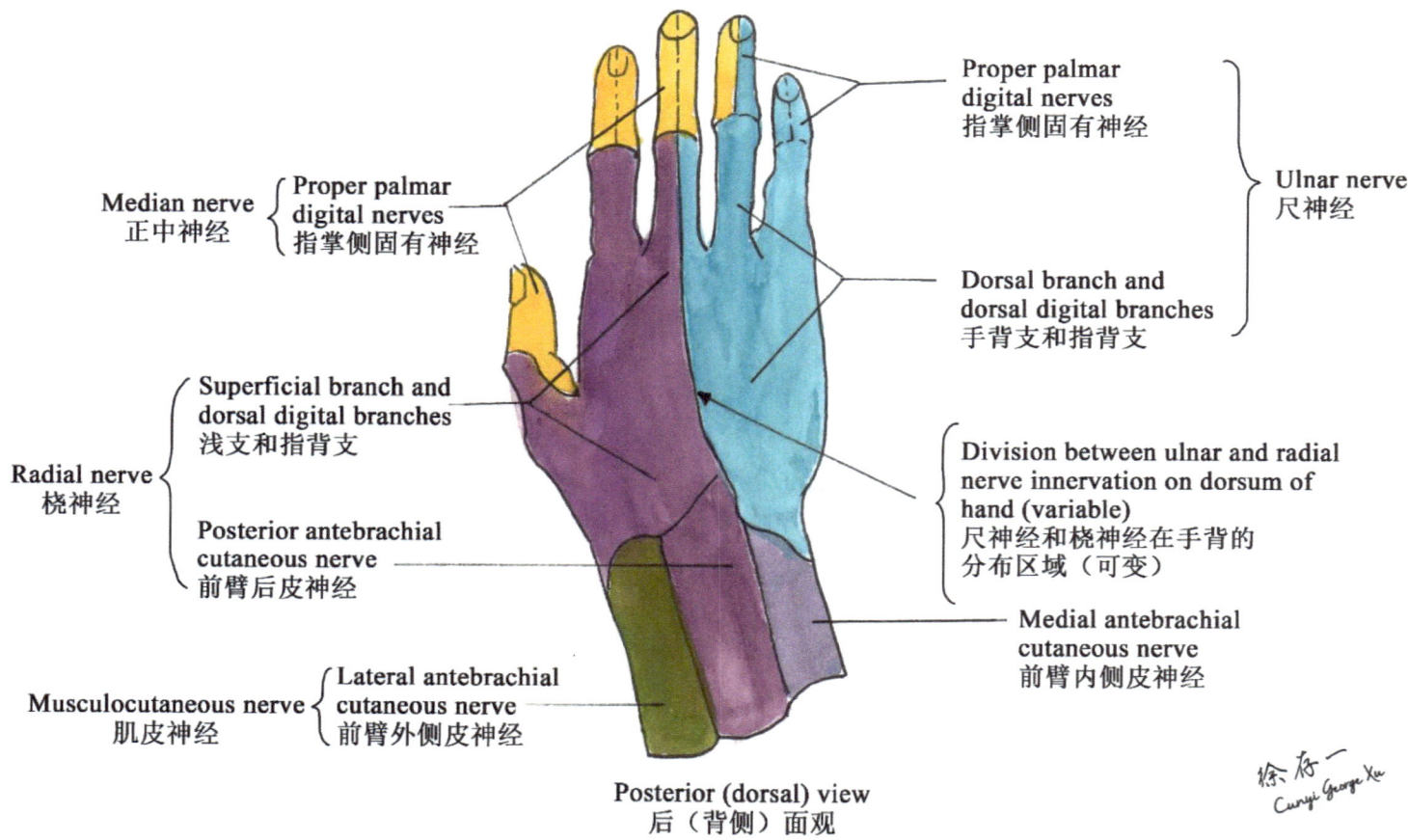

Proper palmar
digital nerves
指掌侧固有神经

Ulnar nerve
尺神经

Median nerve
正中神经

Proper palmar
digital nerves
指掌侧固有神经

Dorsal branch and
dorsal digital branches
手背支和指背支

Superficial branch and
dorsal digital branches
浅支和指背支

Radial nerve
桡神经

Division between ulnar and radial
nerve innervation on dorsum of
hand (variable)
尺神经和桡神经在手背的
分布区域（可变）

Posterior antebrachial
cutaneous nerve
前臂后皮神经

Medial antebrachial
cutaneous nerve
前臂内侧皮神经

Musculocutaneous nerve
肌皮神经

Lateral antebrachial
cutaneous nerve
前臂外侧皮神经

Posterior (dorsal) view
后（背侧）面观

掌（后）神经：手指和手背的感觉与运动。

Palmar (dorsal) nerves: responsible for the sensation and motor function of the fingers and the back of the hand.

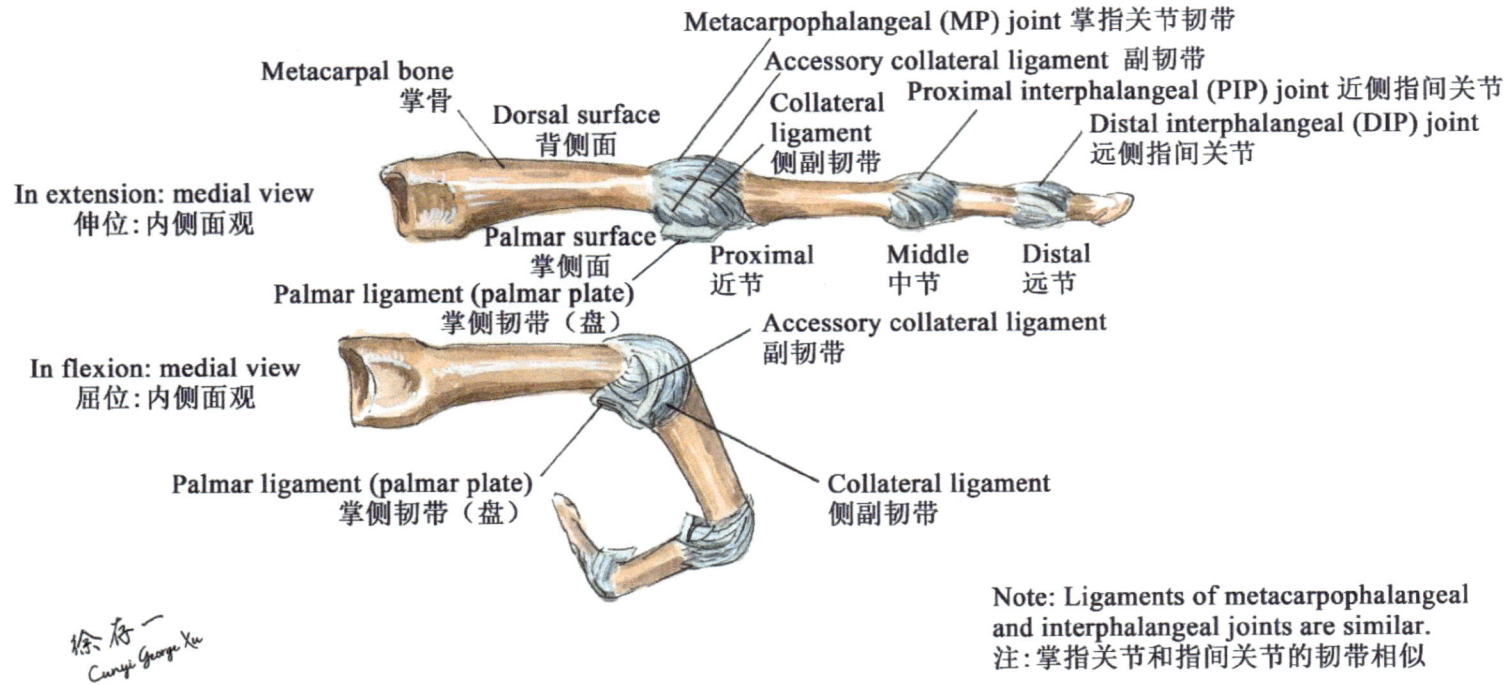

Metacarpophalangeal (MP) joint 掌指关节韧带

Accessory collateral ligament 副韧带

Proximal interphalangeal (PIP) joint 近侧指间关节

Distal interphalangeal (DIP) joint 远侧指间关节

Metacarpal bone 掌骨

Dorsal surface 背侧面

Collateral ligament 侧副韧带

In extension: medial view 伸位：内侧面观

Palmar surface 掌侧面

Proximal 近节

Middle 中节

Distal 远节

Palmar ligament (palmar plate) 掌侧韧带（盘）

In flexion: medial view 屈位：内侧面观

Accessory collateral ligament 副韧带

Palmar ligament (palmar plate) 掌侧韧带（盘）

Collateral ligament 侧副韧带

Cunyi George Xu

Note: Ligaments of metacarpophalangeal and interphalangeal joints are similar.
注：掌指关节和指间关节的韧带相似

掌指韧带和指间韧带：维持手指关节的稳定性和支撑。

Palmaris longus and interosseous ligaments: maintain stability and provide support to the finger joints.

初级人体解剖图谱

Anterior view
前面观

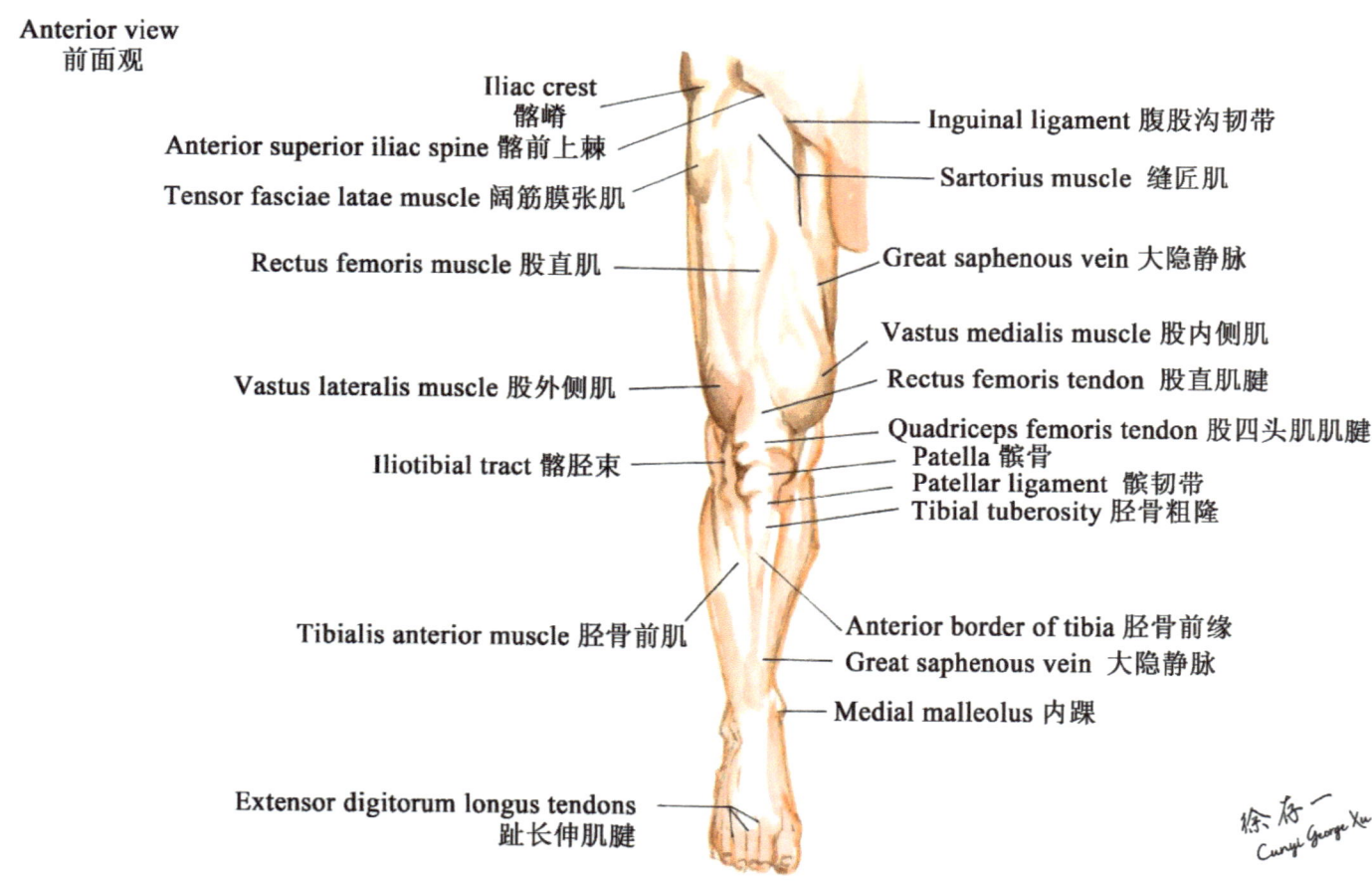

Iliac crest 髂嵴

Anterior superior iliac spine 髂前上棘

Tensor fasciae latae muscle 阔筋膜张肌

Rectus femoris muscle 股直肌

Vastus lateralis muscle 股外侧肌

Iliotibial tract 髂胫束

Tibialis anterior muscle 胫骨前肌

Extensor digitorum longus tendons
趾长伸肌腱

Inguinal ligament 腹股沟韧带

Sartorius muscle 缝匠肌

Great saphenous vein 大隐静脉

Vastus medialis muscle 股内侧肌

Rectus femoris tendon 股直肌腱

Quadriceps femoris tendon 股四头肌肌腱

Patella 髌骨

Patellar ligament 髌韧带

Tibial tuberosity 胫骨粗隆

Anterior border of tibia 胫骨前缘

Great saphenous vein 大隐静脉

Medial malleolus 内踝

Cunyi George Xu

下肢：支撑身体重量、实现行走和奔跑等运动，保持身体的平衡和稳定性。

Lower limbs: support the body's weight, enable movements such as walking and running, and maintain the body's balance and stability.

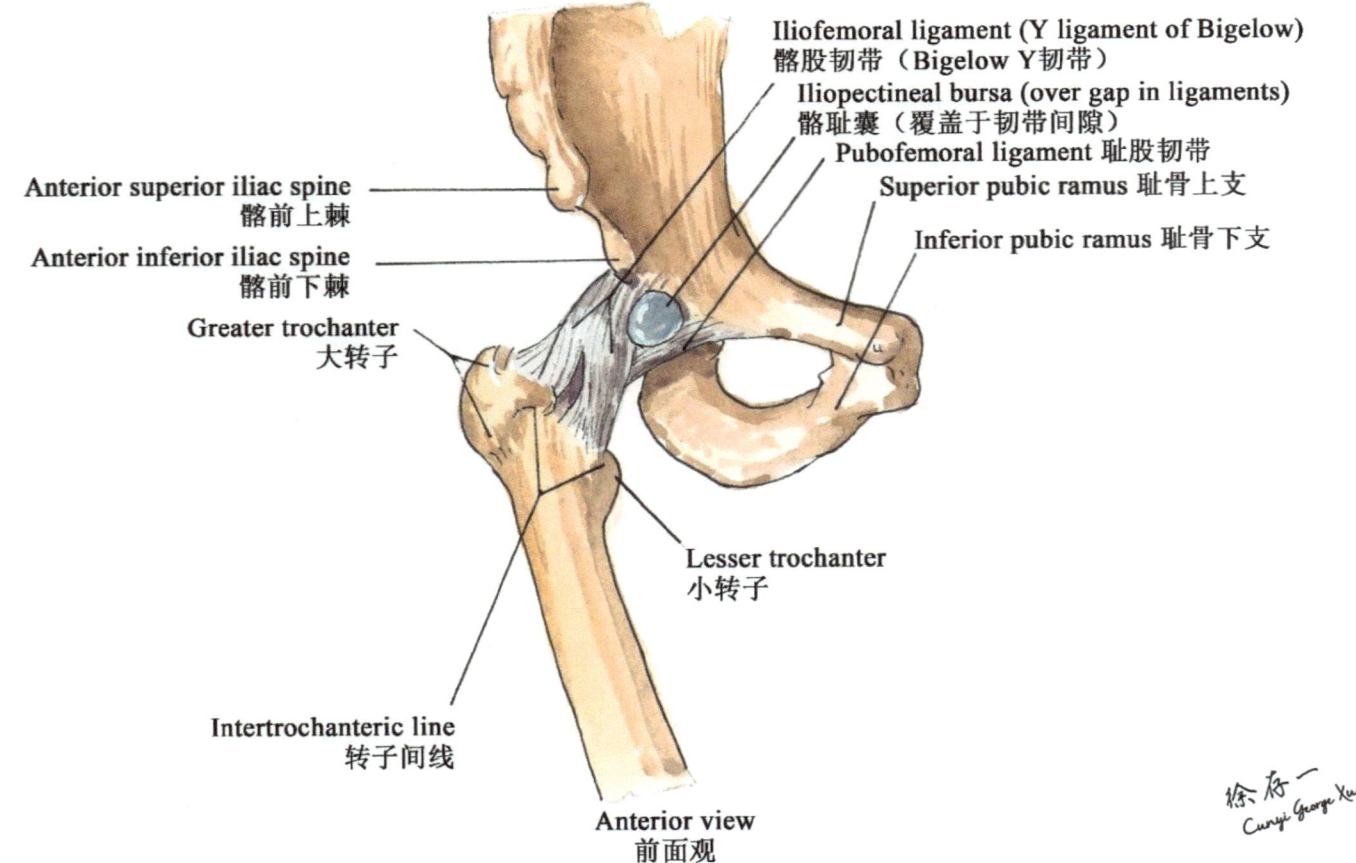

Iliofemoral ligament (Y ligament of Bigelow)
髂股韧带（Bigelow Y韧带）
Iliopectineal bursa (over gap in ligaments)
髂耻囊（覆盖于韧带间隙）
Pubofemoral ligament 耻股韧带
Superior pubic ramus 耻骨上支
Inferior pubic ramus 耻骨下支

Anterior superior iliac spine
髂前上棘
Anterior inferior iliac spine
髂前下棘
Greater trochanter
大转子
Lesser trochanter
小转子
Intertrochanteric line
转子间线

Anterior view
前面观

徐存一
Cunyi George Xu

髋关节：支撑体重并允许进行广泛的运动，包括行走、跑步、爬楼梯以及扭转和弯腰等活动。

Hip joint: supports body weight and allows for a wide range of motion, including walking, running, climbing stairs, and activities such as twisting and bending over.

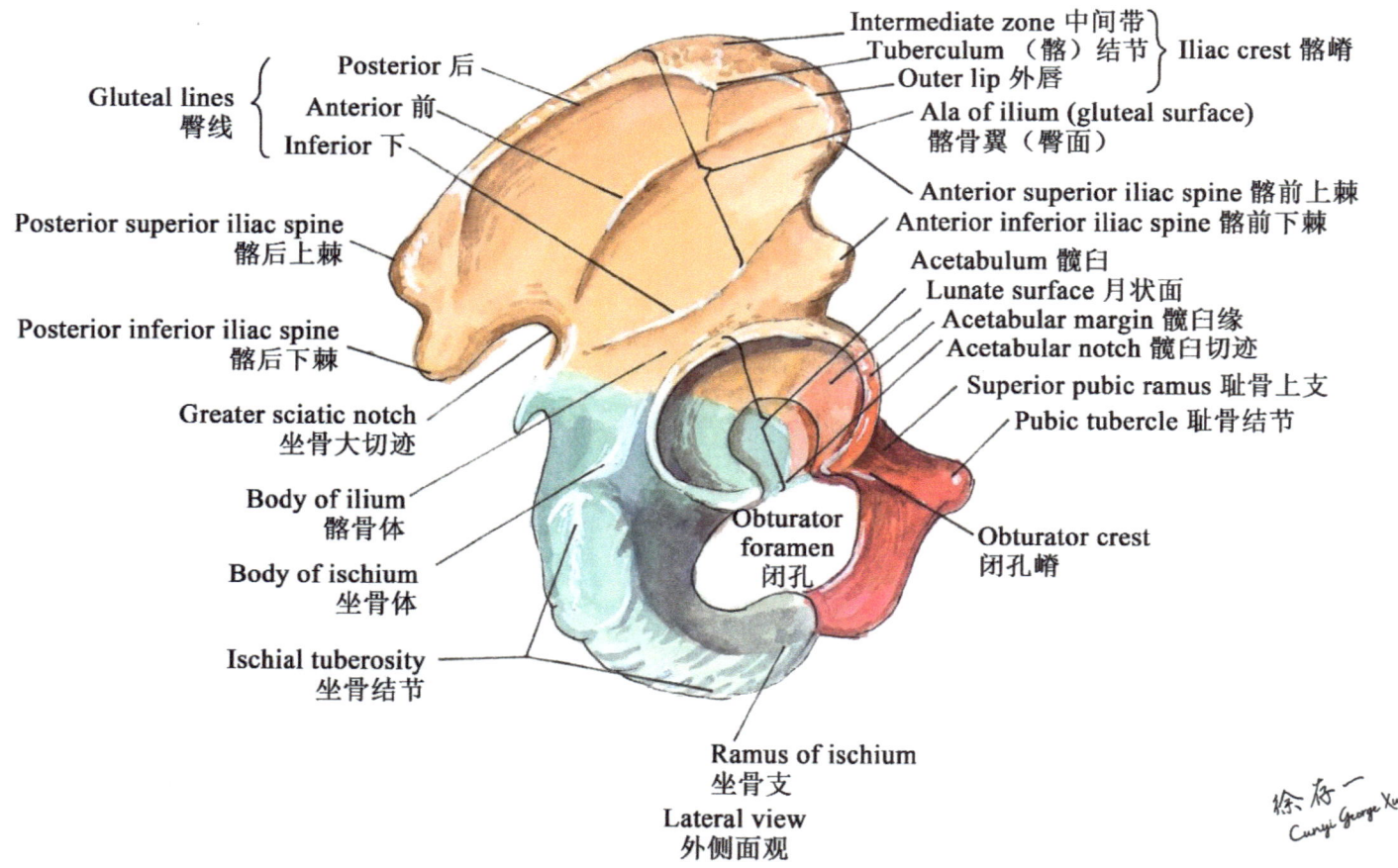

Gluteal lines 臀线 { Posterior 后 / Anterior 前 / Inferior 下

Intermediate zone 中间带 } Iliac crest 髂嵴
Tuberculum （髂）结节
Outer lip 外唇

Ala of ilium (gluteal surface) 髂骨翼（臀面）

Anterior superior iliac spine 髂前上棘
Anterior inferior iliac spine 髂前下棘

Posterior superior iliac spine 髂后上棘

Posterior inferior iliac spine 髂后下棘

Acetabulum 髋臼
Lunate surface 月状面
Acetabular margin 髋臼缘
Acetabular notch 髋臼切迹

Superior pubic ramus 耻骨上支
Pubic tubercle 耻骨结节

Greater sciatic notch 坐骨大切迹

Body of ilium 髂骨体

Body of ischium 坐骨体

Obturator foramen 闭孔

Obturator crest 闭孔嵴

Ischial tuberosity 坐骨结节

Ramus of ischium 坐骨支

Lateral view 外侧面观

徐存一 Cunyi George Xu

髋骨：支撑上半身，连接脊柱和下肢，同时为下肢肌肉提供牢固的附着点，并在行走和保持平衡时发挥关键作用。

Hip bone: supports the upper body, connects the spine to the lower limbs, provides a sturdy attachment point for the muscles of the lower limbs, and plays a key role in walking and maintaining balance.

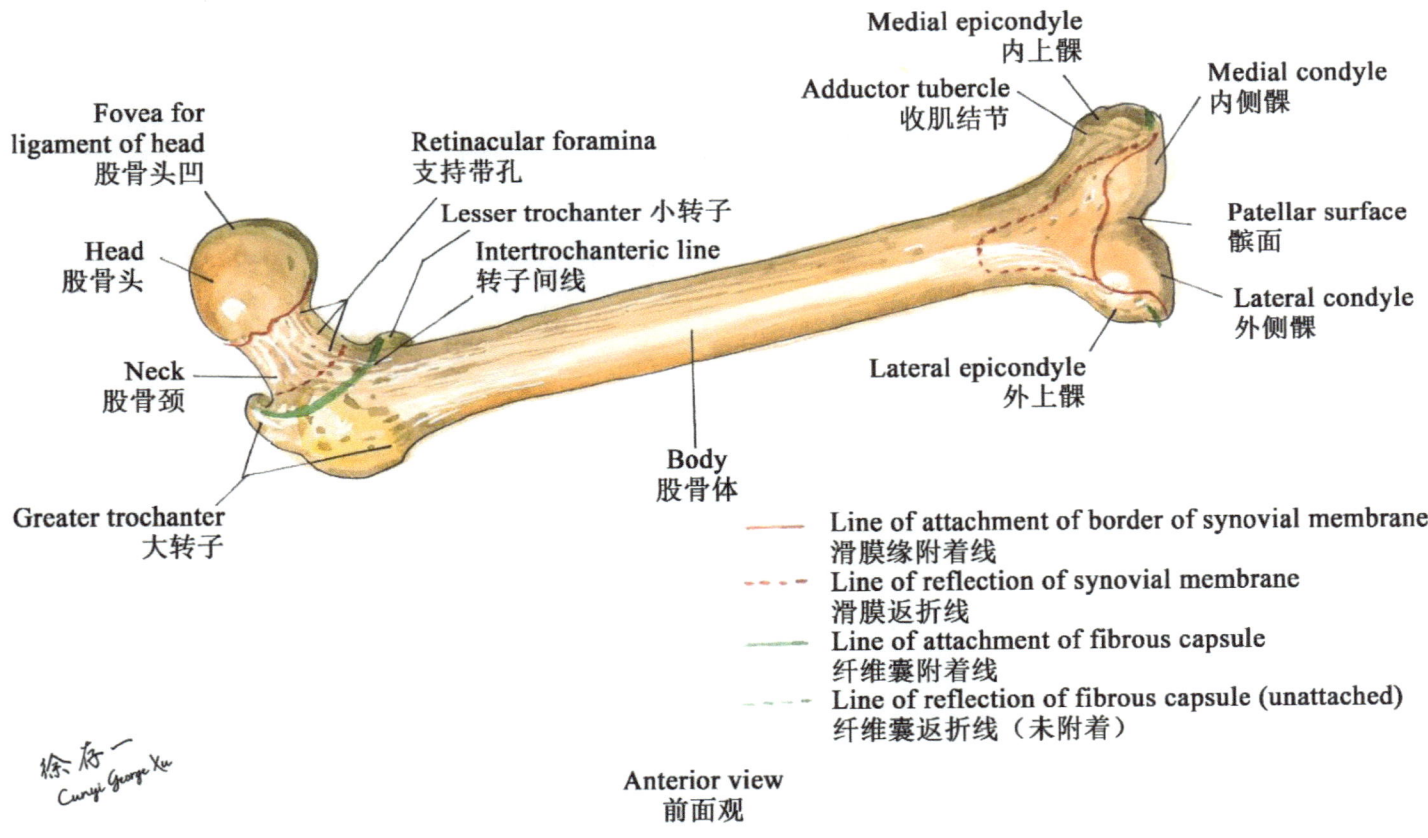

Fovea for
ligament of head
股骨头凹

Head
股骨头

Neck
股骨颈

Greater trochanter
大转子

Retinacular foramina
支持带孔

Lesser trochanter 小转子

Intertrochanteric line
转子间线

Body
股骨体

Medial epicondyle
内上髁

Adductor tubercle
收肌结节

Medial condyle
内侧髁

Patellar surface
髌面

Lateral condyle
外侧髁

Lateral epicondyle
外上髁

—— Line of attachment of border of synovial membrane
滑膜缘附着线
- - - Line of reflection of synovial membrane
滑膜返折线
—— Line of attachment of fibrous capsule
纤维囊附着线
- · - Line of reflection of fibrous capsule (unattached)
纤维囊返折线（未附着）

徐存一
Cunyi George Xu

Anterior view
前面观

股骨：支撑上半身的重量、提供运动时的杠杆作用，与髋关节和膝关节协同工作以实现行走、跑步和其他下肢活动。

Femur: supports the weight of the upper body, provides leverage during exercise, and works in conjunction with the hip and knee joints to facilitate walking, running, and other lower limb activities.

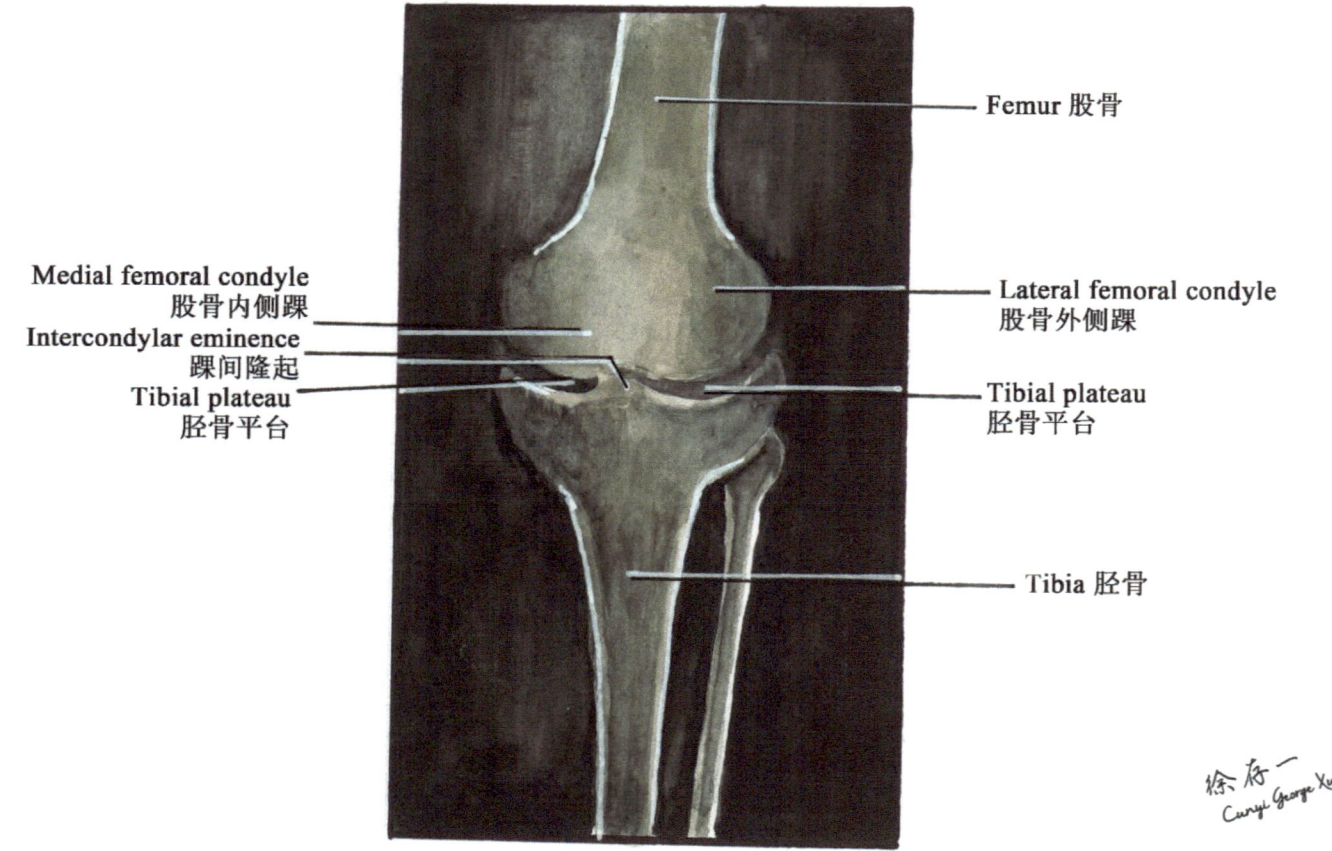

Femur 股骨

Medial femoral condyle
股骨内侧踝

Intercondylar eminence
踝间隆起

Tibial plateau
胫骨平台

Lateral femoral condyle
股骨外侧踝

Tibial plateau
胫骨平台

Tibia 胫骨

膝：稳定膝关节，维持运动功能，支持人体的站立和行走。

Knee: stabilises the knee joint, maintains motor function, and supports the body's ability to stand and walk.

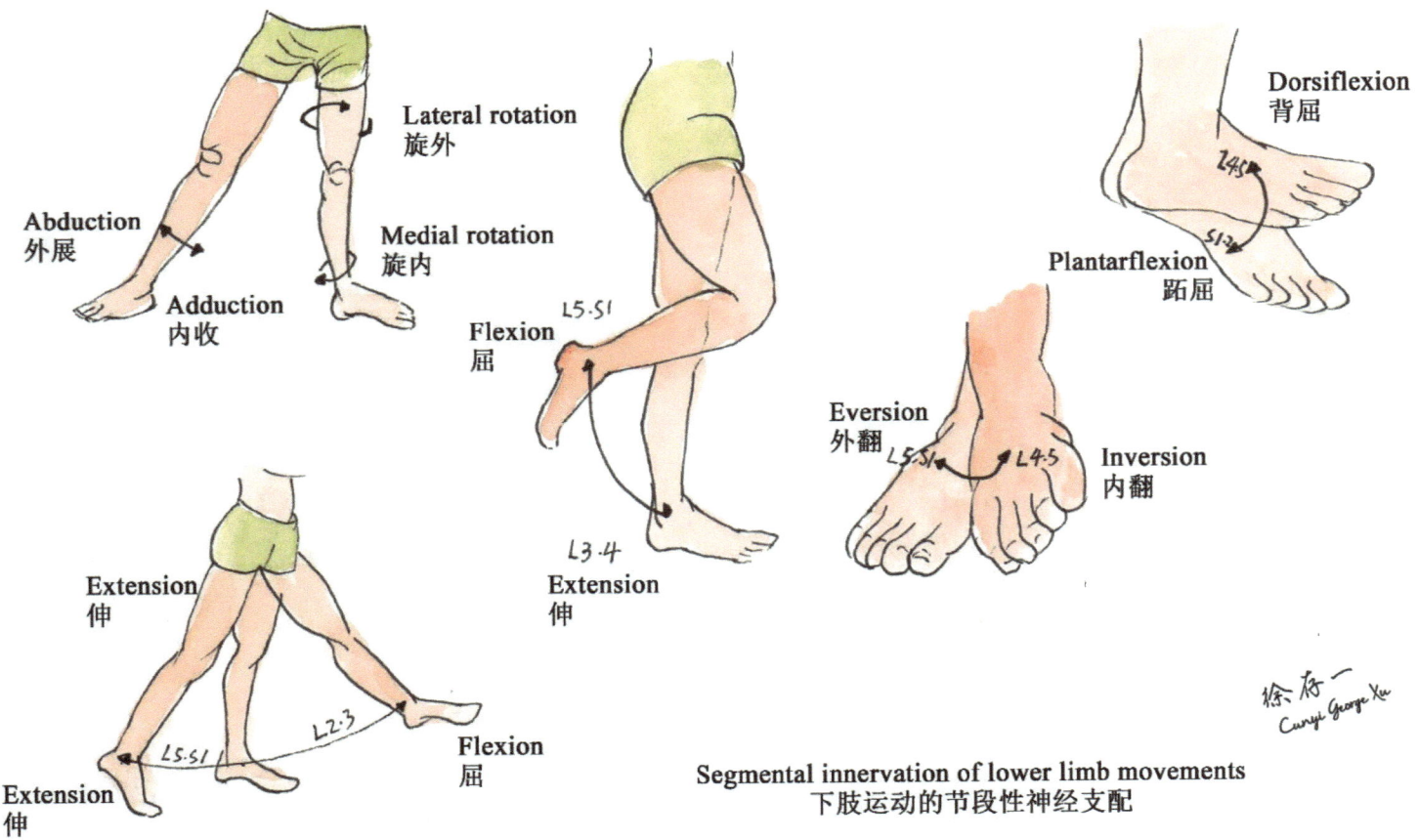

Lateral rotation
旋外

Medial rotation
旋内

Abduction
外展

Adduction
内收

Flexion
屈

L5-S1

L3-4

Extension
伸

Dorsiflexion
背屈

L4-5

Plantarflexion
跖屈

S1-2

Eversion
外翻   L5-S1

L4-5   Inversion
内翻

Extension
伸

Flexion
屈

L5-S1

L2-3

Extension
伸

徐存一
Cunyi George Xu

Segmental innervation of lower limb movements
下肢运动的节段性神经支配

下肢运动的节段性神经支配：调节肌肉运动，使身体能够正常运动，并维持身体的平衡。

Segmental innervation of lower limb movements: regulates muscle movement to enable the body to move normally and maintain balance.

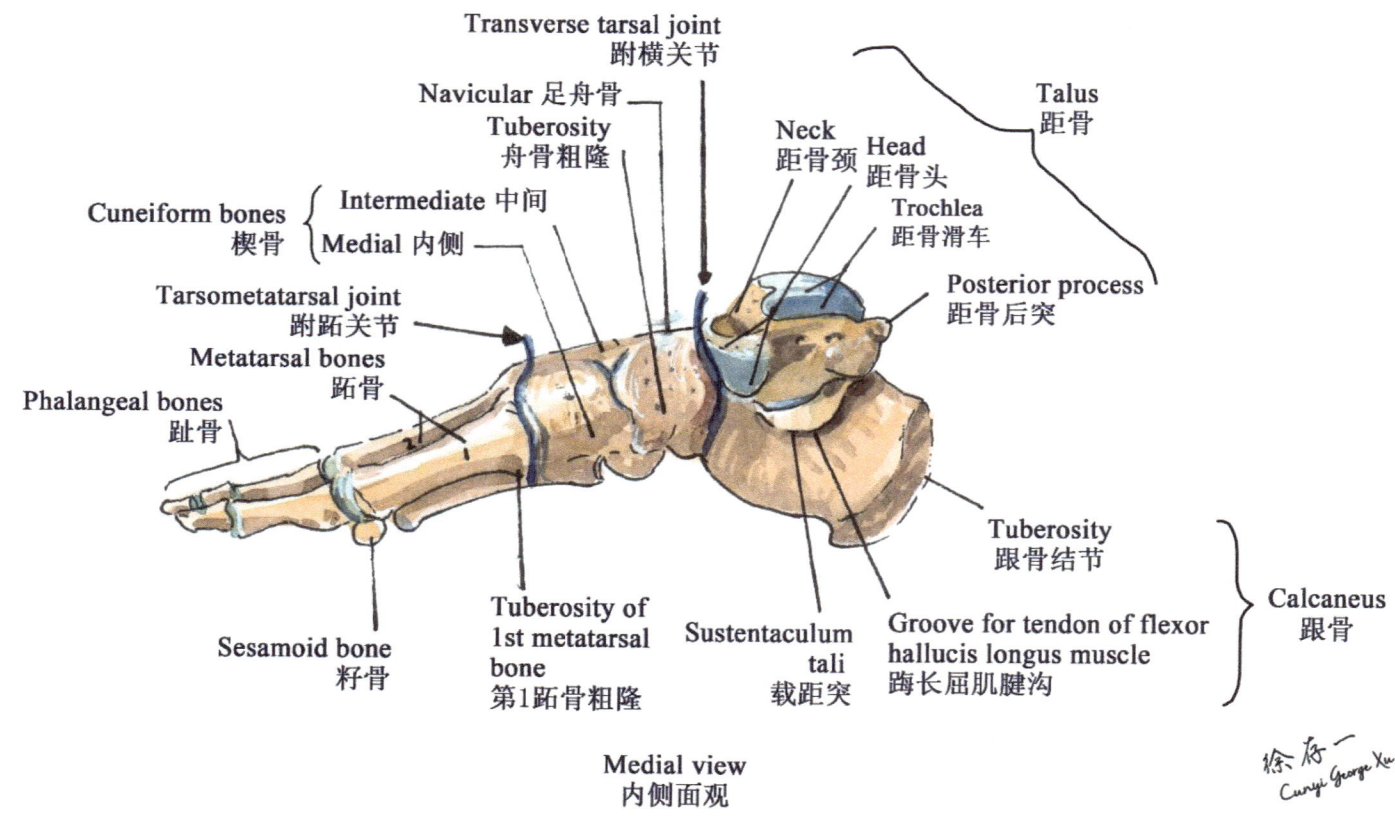

Transverse tarsal joint
跗横关节

Navicular 足舟骨

Tuberosity
舟骨粗隆

Cuneiform bones
楔骨

Intermediate 中间

Medial 内侧

Tarsometatarsal joint
跗跖关节

Metatarsal bones
跖骨

Phalangeal bones
趾骨

Sesamoid bone
籽骨

Tuberosity of
1st metatarsal
bone
第1跖骨粗隆

Neck
距骨颈

Head
距骨头

Trochlea
距骨滑车

Talus
距骨

Posterior process
距骨后突

Tuberosity
跟骨结节

Calcaneus
跟骨

Sustentaculum
tali
载距突

Groove for tendon of flexor
hallucis longus muscle
踇长屈肌腱沟

Medial view
内侧面观

足骨：支持体重、维持姿势与参与下肢运动。

Bone of the foot: supports body weight, maintains posture, and participates in lower limb movements.

第六章　微观世界—细胞与病毒

Chapter 6 Micro World—Cells and Viruses

初级人体解剖图谱

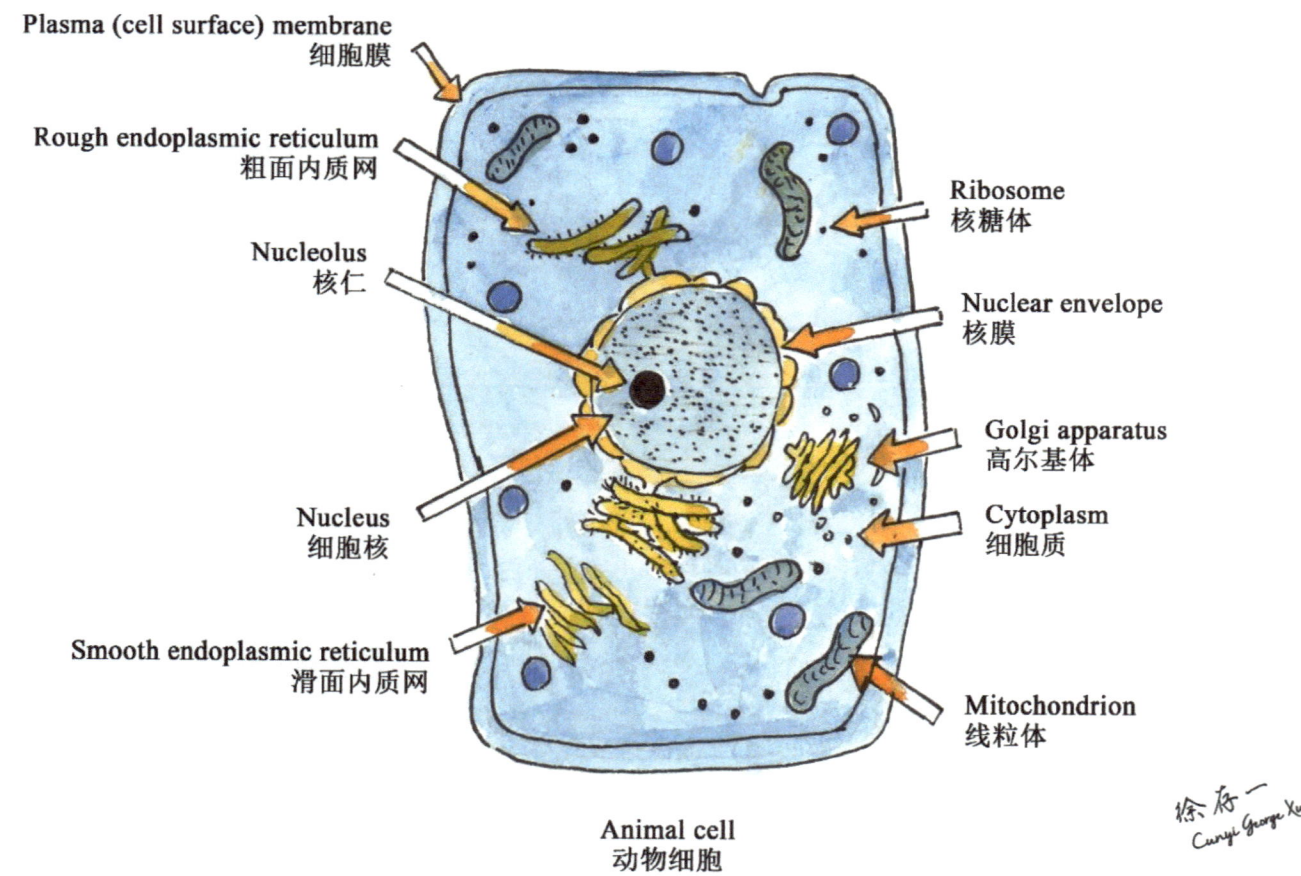

Plasma (cell surface) membrane
细胞膜

Rough endoplasmic reticulum
粗面内质网

Nucleolus
核仁

Nucleus
细胞核

Smooth endoplasmic reticulum
滑面内质网

Ribosome
核糖体

Nuclear envelope
核膜

Golgi apparatus
高尔基体

Cytoplasm
细胞质

Mitochondrion
线粒体

Animal cell
动物细胞

动物细胞： 构成各种器官进行生命活动，完成动物的生理功能。

Animal cell: forms various organs for life activities and performs physiological functions in animals.

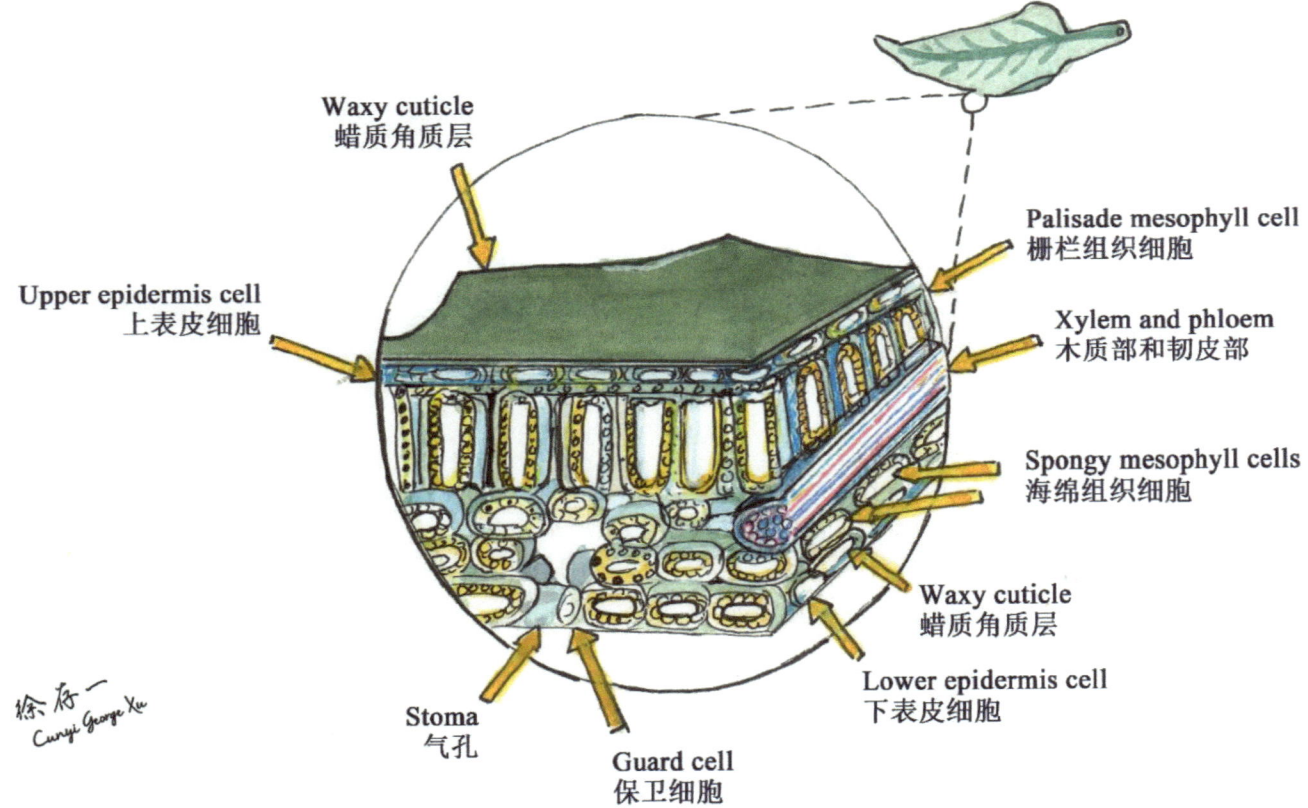

Waxy cuticle
蜡质角质层

Upper epidermis cell
上表皮细胞

Palisade mesophyll cell
栅栏组织细胞

Xylem and phloem
木质部和韧皮部

Spongy mesophyll cells
海绵组织细胞

Waxy cuticle
蜡质角质层

Lower epidermis cell
下表皮细胞

Stoma
气孔

Guard cell
保卫细胞

徐存一
Cunyi George Xu

植物叶片：叶的主体部分，进行光合作用和蒸腾作用。

Leaf: the main part of a plant that performs photosynthesis and transpiration.

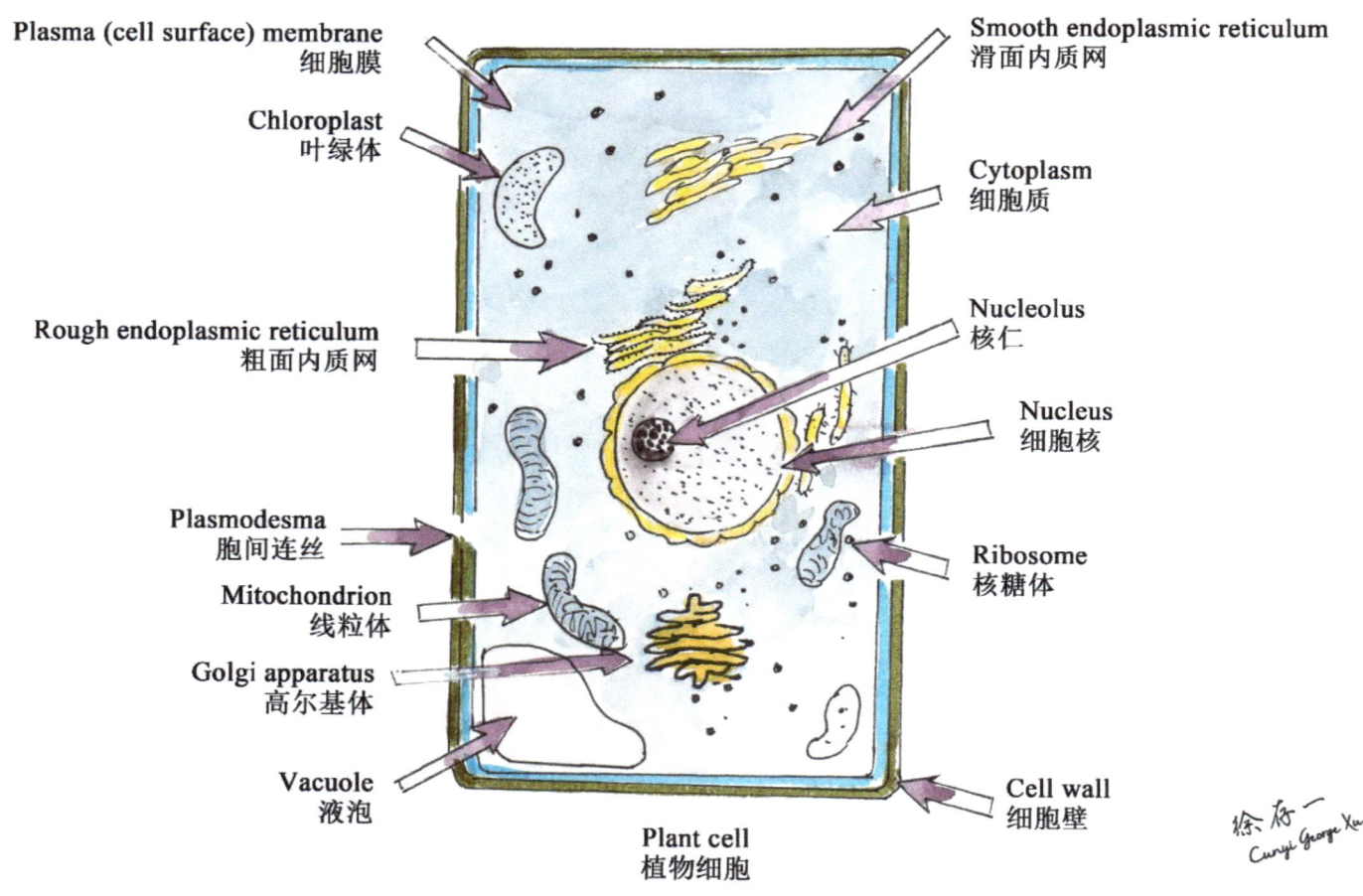

Plasma (cell surface) membrane
细胞膜

Chloroplast
叶绿体

Rough endoplasmic reticulum
粗面内质网

Plasmodesma
胞间连丝

Mitochondrion
线粒体

Golgi apparatus
高尔基体

Vacuole
液泡

Smooth endoplasmic reticulum
滑面内质网

Cytoplasm
细胞质

Nucleolus
核仁

Nucleus
细胞核

Ribosome
核糖体

Cell wall
细胞壁

Plant cell
植物细胞

植物细胞：维持细胞的形态、保护细胞内部结构、进行物质交换和能量转换。

Plant cell: maintains cell morphology, protects internal structure, and facilitates material exchange and energy conversion.

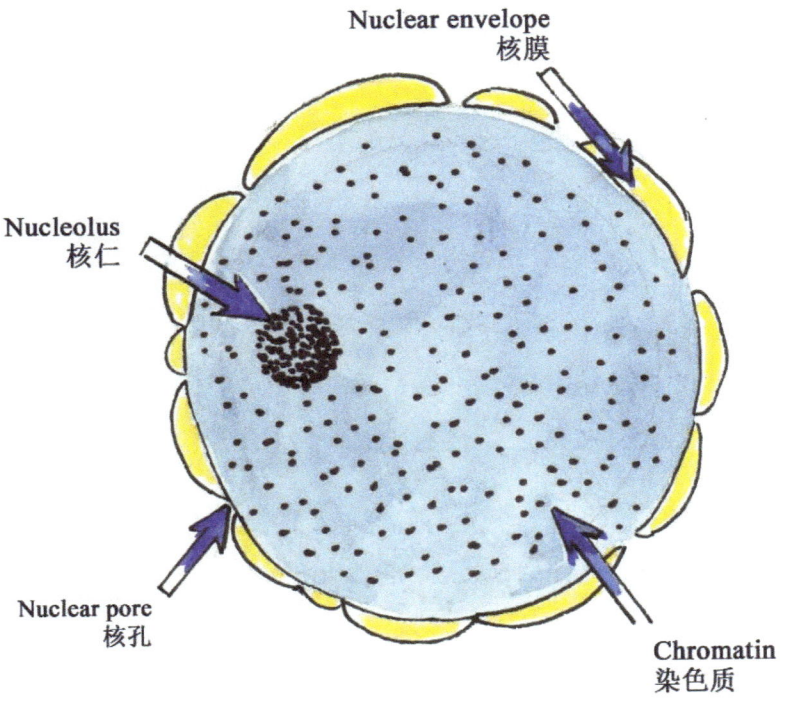

Nuclear envelope
核膜

Nucleolus
核仁

Nuclear pore
核孔

Chromatin
染色质

Nucleus
细胞核

细胞核：细胞的遗传和代谢控制中心。

Nucleus: the genetic and metabolic control centre of the cell.

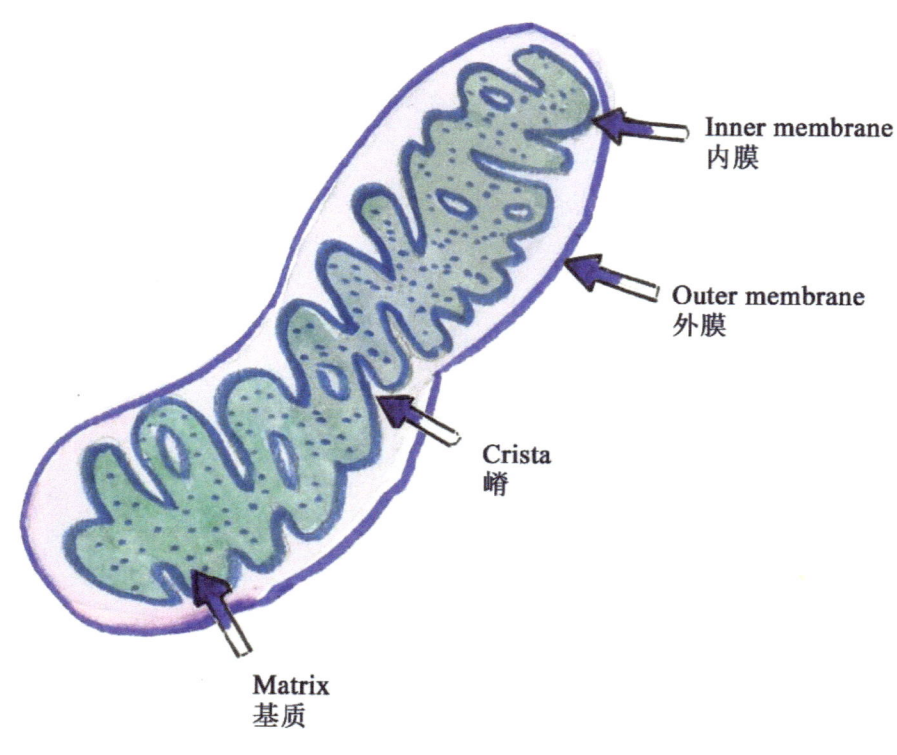

Inner membrane
内膜

Outer membrane
外膜

Crista
嵴

Matrix
基质

Mitochondrion
线粒体

线粒体：细胞能量代谢，是细胞进行有氧呼吸的主要场所。

Mitochondria: responsible for cellular energy metabolism and aerobic respiration.

Vesicle
囊泡

Golgi apparatus
高尔基体

高尔基体：完成细胞分泌物（如蛋白）最后加工和包装的场所。

Golgi Apparatus: completes the final processing and packaging of cellular secretions (such as proteins).

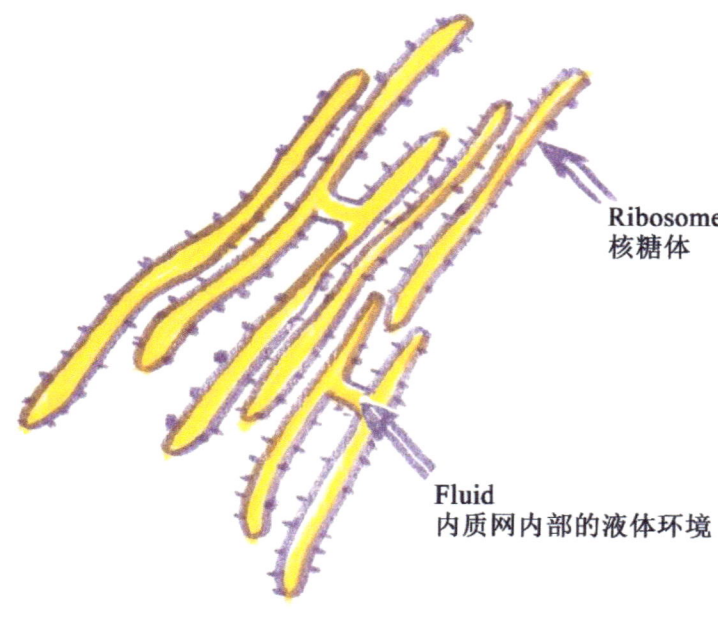

Ribosome
核糖体

Fluid
内质网内部的液体环境

Rough endoplasmic reticulum (RER)
粗面内质网

粗面内质网：合成蛋白质大分子，并从细胞输送出去或在细胞内转运到其他部位。

Rough Endoplasmic Reticulum: synthesises protein macromolecules and transports them out of the cell or to other parts of the cell.

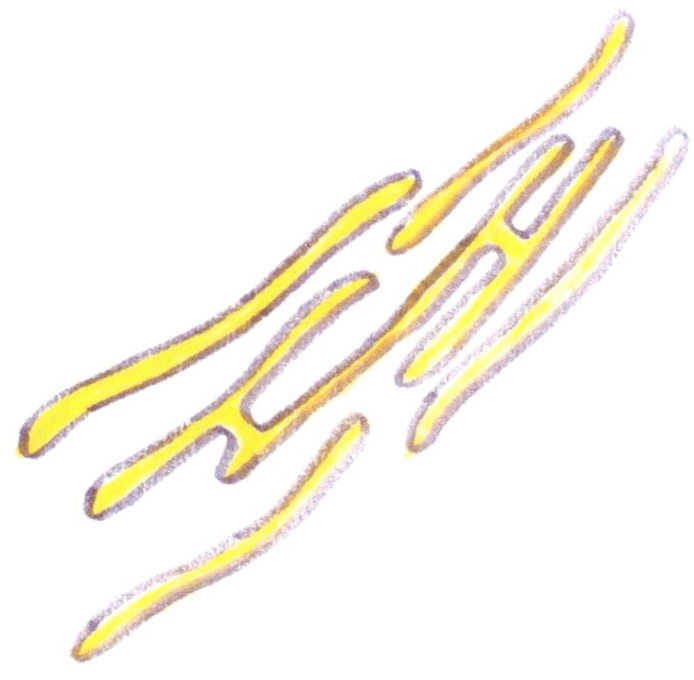

Smooth endoplasmic reticulum (SER)
滑面内质网

滑面内质网：与糖类和脂类的合成、解毒、同化作用有关，还具有运输蛋白质的功能。

Smooth Endoplasmic Reticulum: involved in the synthesis, detoxification, and assimilation of sugars and lipids, and has the function of transporting proteins.

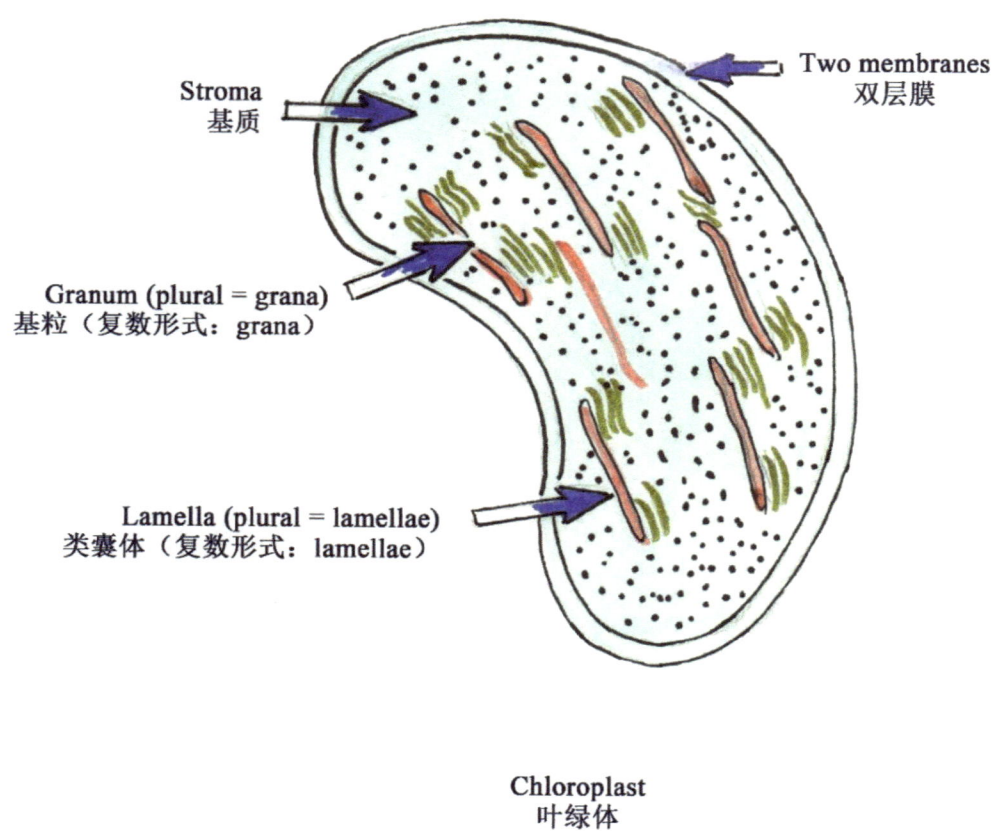

Stroma
基质

Two membranes
双层膜

Granum (plural = grana)
基粒（复数形式：grana）

Lamella (plural = lamellae)
类囊体（复数形式：lamellae）

Chloroplast
叶绿体

叶绿体：进行光合作用，合成有机物质并释放氧气。

Chloroplast: responsible for photosynthesis, the synthesis of organic matter, and the release of oxygen.

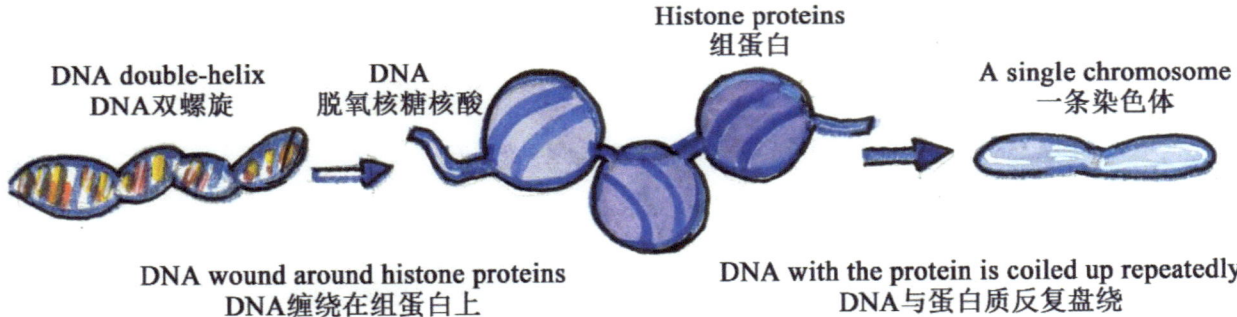

DNA double-helix
DNA双螺旋

DNA
脱氧核糖核酸

Histone proteins
组蛋白

A single chromosome
一条染色体

DNA wound around histone proteins
DNA缠绕在组蛋白上

DNA with the protein is coiled up repeatedly
DNA与蛋白质反复盘绕

染色体组装过程：DNA 与组蛋白结合形成核小体，进一步组装成长染色体。

Chromosome assembly: DNA binds with histones to form nucleosomes, which are further assembled into long chromosomes.

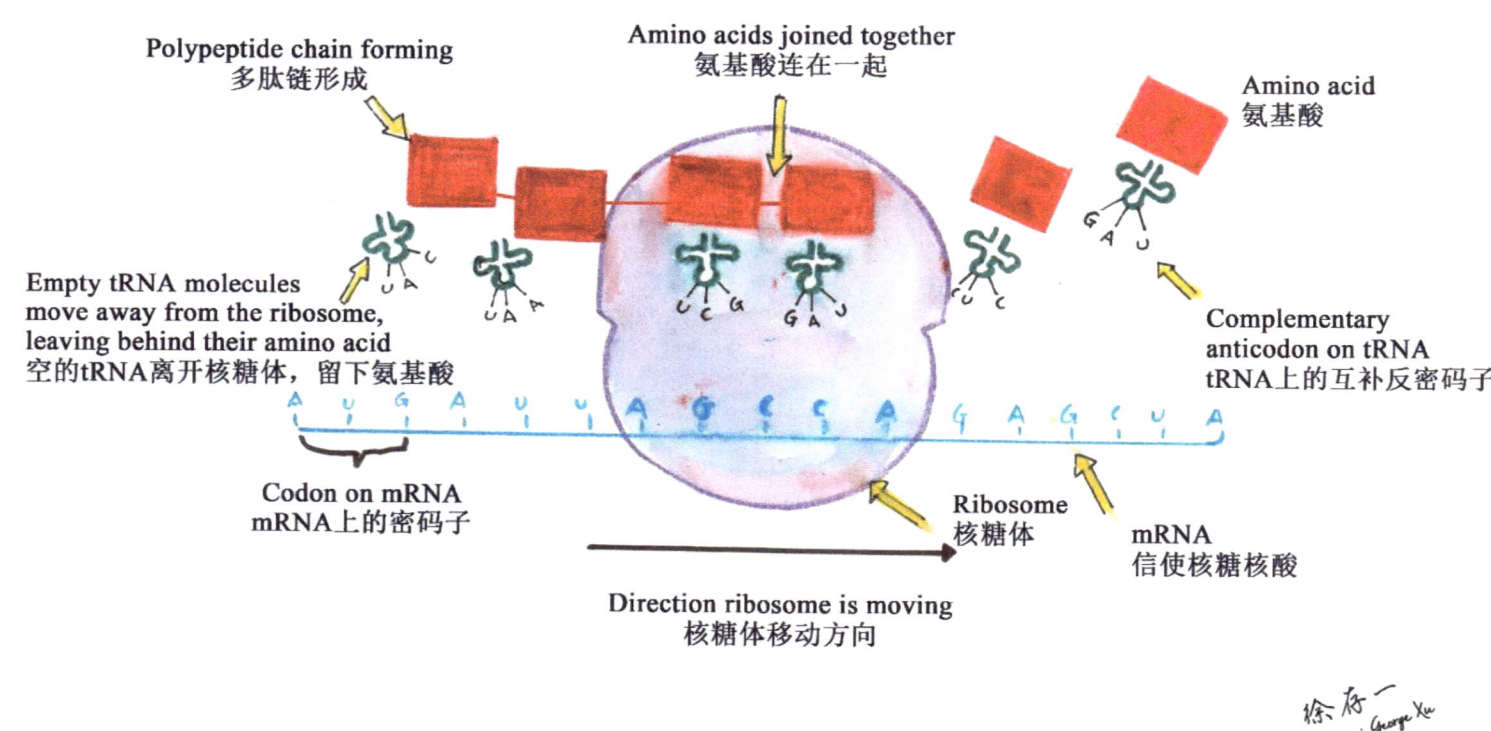

Polypeptide chain forming
多肽链形成

Amino acids joined together
氨基酸连在一起

Amino acid
氨基酸

Empty tRNA molecules move away from the ribosome, leaving behind their amino acid
空的tRNA离开核糖体，留下氨基酸

Complementary anticodon on tRNA
tRNA上的互补反密码子

Codon on mRNA
mRNA上的密码子

Ribosome
核糖体

mRNA
信使核糖核酸

Direction ribosome is moving
核糖体移动方向

蛋白质翻译过程：细胞利用 mRNA 上的遗传信息，在核糖体中合成具有特定氨基酸序列的多肽链的过程。

Protein translation: Cells utilise the genetic information on mRNA to synthesise polypeptide chains with specific sequences of amino acids in the ribosome.

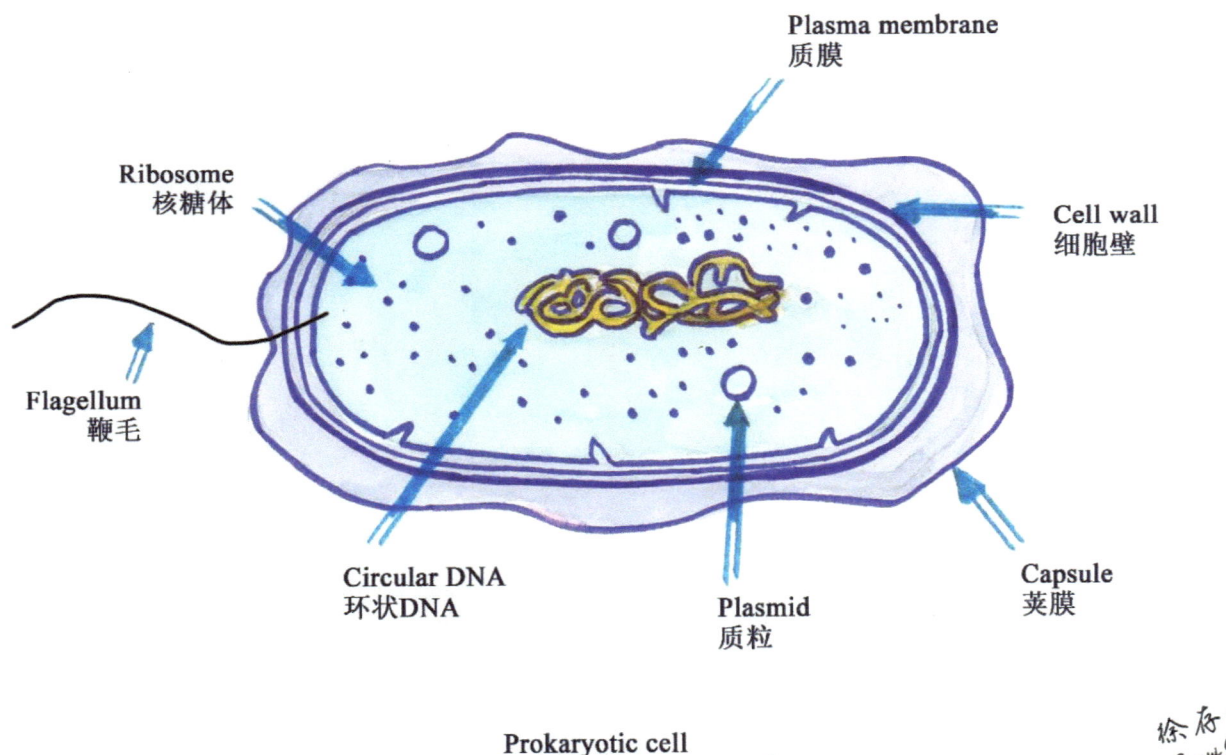

Plasma membrane
质膜

Ribosome
核糖体

Cell wall
细胞壁

Flagellum
鞭毛

Circular DNA
环状DNA

Plasmid
质粒

Capsule
荚膜

Prokaryotic cell
原核细胞

Cunyi George Xu

原核细胞：主要特征是没有以核膜为界的细胞核。

Prokaryotic cell: Characterised by the absence of a nucleus bounded by a nuclear membrane.

Viruses contain a core of genetic material — either DNA or RNA.
病毒含有一个遗传物质核心——DNA或RNA。

The protein coat around the core is called the capsid.
核心周围的蛋白质外壳被称为衣壳。

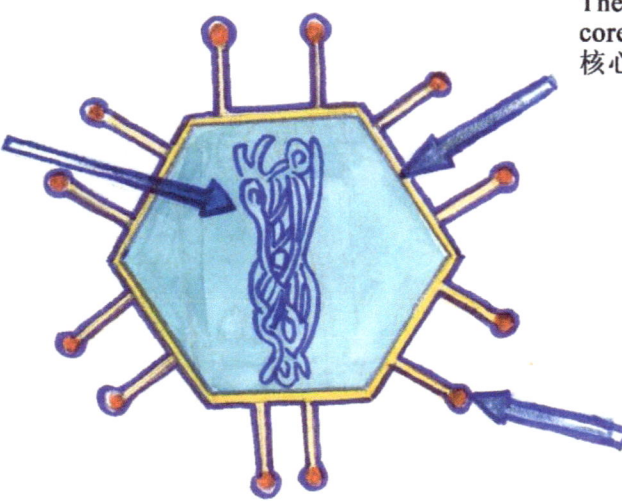

Attachment proteins stick out from the edge of the capsid. These let the virus cling on to a suitable host cell.
附着蛋白从衣壳边缘伸出来。这些蛋白让病毒能够附着在合适的宿主细胞上。

Viruses
病毒

病毒：非细胞生命形态，在活细胞内寄生并以复制方式增殖。

Virus: Acellular, parasitic within living cells, and proliferating through replication.

www.ingramcontent.com/pod-product-compliance
Lightning Source LLC
Chambersburg PA
CBHW040812120626
46547CB00004B/527